14th International Conference on Applied Physics and Mathematics 2024

Journal of Physics: Conference Series Volume 2793

Taipei, Taiwan
10-12 April 2024

ISBN: 979-8-3313-0207-8
ISSN: 1742-6588

Printed from e-media with permission by:

Curran Associates, Inc.
57 Morehouse Lane
Red Hook, NY 12571

Some format issues inherent in the e-media version may also appear in this print version.

Institute of Physics
Dirac House, Temple Back
Bristol BS1 6BE UK

Phone: 44 1 17 929 7481
Fax: 44 1 17 920 0979

techtracking@iop.org

Additional copies of this publication are available from:

Curran Associates, Inc.
57 Morehouse Lane
Red Hook, NY 12571 USA
Phone: 845-758-0400
Fax: 845-758-2633
Email: curran@proceedings.com
Web: www.proceedings.com

TABLE OF CONTENTS

Preface

Peer Review Statement

Numerical Analysis Clarifying Non-Equilibrated Periodic Flows Generated by Dual-Scaling Taylor's Analytical Solution 1
Xuanyou Gong, Hiroki Suzuki, Toshinori Kouchi

Can Newtonian Kinetic Energy and Einsteinian Rest-Mass Energy Be Expressed by the Binomial Expansion of the Lorentz Factor? and How Valid is Using Einstein's E = Mc2 to Calculate the Nuclear Fission Energy? 7
Talal Al-Ameen, Maythem Mahmud, Imad Muhi

Investigation of the Weight Effect on the Oscillations of a Magnetically-Levitated Body Coupled to an Energy Harvester 14
A C Francis, A Kandil, A Elsaid, W K Zahra

Attempt to Compensate for Effects of Energy Conservation Error on Static Pressure Fluctuation Using Sub-Grid Scale Components with the Reynolds Number Dependence of Isotropic/Anisotropic Steady Turbulence 21
Makoto Chitose, Hiroki Suzuki, Toshinori Kouchi

Analytical Fermi Energy for Broadened Landau Levels of 2DEG Under a Perpendicular Strong Magnetic Field 27
L J Sese, C J Bercasio, R Gammag

Heating Opaque Material Sheets by a Blue Laser 32
Kwanruthai Sibsiri, Puchong Kijamnajsuk, Malee Sutthiopad, Chaiya Luengviriya

Ab Initio Study of Gallium-Doped Zinc Oxides and Zinc-Doped Gallium Oxides 37
Cheng-Lung Yu, Jian-Sheng Xie Su, Hsiang-Yu Hsieh, Po-Liang Liu

Predicted Yield and Detectability of GW Events from the IllustrisTNG Simulations 43
Luchang Niu, Tiangang Zhang, Kun Zhang, Yiting Li

On Propagation of Gradual and Impulsive Coronal Mass Ejections (CMEs) 52
N Mohamad Ansor, Z S Hamidi, N N M Shariff

Multi-Objective Optimal Design for the Railway Bogie Frame System Under Stress and Fatigue Analysis 60
Cheng-Kang Lee, Chien-Kuan Liu, Yung-Chang Cheng

Preliminary Analysis for a New Sub-Type III Solar Radio Burst (Type VI) on 11th July 2023 66
N. H. M. Bokhari, Z. S. Hamidi, N. N. M. Shariff

Linear Intersection Pairs of Negacyclic Codes and Their Applications in Construction of Entanglement-Assisted Quantum Error-Correcting Codes 73
Somphong Jitman

Effects of Strong Solar Flares on the Ionospheric Electron Density Over the Philippines 79
Zane Nikia C. Domingo, Ernest P. Macalalad

A Reliable Semi-Analytic Solution for the Caputo Fractional Human Liver Model 88
Abrar Ansar, Marwan Abukhaled

Dynamical Concurrence of Initially Separable Bipartite System Immersed in N-Qubit Environment 96
R Alporha, L J Sese, R Gammag

A Survey on Topological Solitons in Planar Nonlinear Dirac Models 102
R Dikole

Predicting the Minimum Energy Pathway of 1H to 1T Phase Transition of Select 2D Transition Metal Dichalcogenides Via Density Functional Theory and Machine Learning Approach 109
Ky M Diego, D B Putungan, A B Santos-Putungan

Detection of Microplastics in Surface Water Using a DSLR Lens-Based UV Imaging System 115
Ra J Jaruda, Jl F Gabayno

Investigation of a MEMS Resonator Model with Quintic Nonlinearity 127
M. E. Abdelraouf, A. Kandil, W. K. Zahra, A. Elsaid

A Monte Carlo Study of K Escape Phenomena of X-Ray Spectra in CdTe and Gd2O2S Detectors 134
Dan Wang, Zhe Shi, Zhi Yang

Author Index

Journal of Physics: Conference Series **2793** (2024) 011001 doi:10.1088/1742-6596/2793/1/011001

Preface

2024 14th International Conference on Applied Physics and Mathematics (ICAPM 2024) was held in Taipei, Taiwan during April 10-12, 2024. In the past 13 years, ICAPM was held in large capital cities, such as Tokyo, Phuket, Singapore, Busan, Lisbon, and Hong Kong, Bangkok, etc. This conference provides a remarkable opportunity for the academic and industrial communities to address new challenges and share solutions, and discuss future research directions.

The conference has lasted for three days. Three keynote speakers were invited to give keynote speech. They are Prof. Naoyuki Ishimura, Chuo University, Japan; Prof. Ying-Hao Chu, National Tsing Hua University, Taiwan and Prof. Hsi-Sheng Goan, National Taiwan University, Taiwan. Scholars and researchers from China, Malaysia, Thailand, Indonesia, Philippines, Russia, and so on took this opportunity to share their research results and discuss the future development of related fields.

All papers in the conference proceedings have passed vigorous preliminary review and peer review by the technical committee. Professional comments are given according to the aspects of originality, innovation, applicability, technical merit, organizing and writing, relevance to conference, and so on. Authors benefit from the valuable comments and improve their submissions to meet the satisfaction of the reviewers. The proceedings contain a collection of research papers, and are divided into four chapters: Chapter one is entitled B Basic Physics Theory and Methods; Chapter two is entitled Astrophysics and Motion Physics; Chapter three is entitled Mathematical Theory and Engineering Mathematics; Chapter four is entitled Mathematical Models and Calculations in Physics.

We would like to especially thank the members of the Technical Committees. They worked very hard to review papers and made valuable suggestions for authors to improve their work. We would also like to thank the external reviewers for their assistance during the review process, and the authors for their contributions to the conference. Without such support, conferences cannot maintain quality or increase truly original contributions to knowledge. It's widely recognized that this is the main reason for its success, but we want to emphasize it again.

On behalf of organizing committee, we would like to express our sincere gratitude to all participants and speakers who contributed to make this event a great success.

Po-Liang Liu
Conference Chair
ICAPM 2023

Journal of Physics: Conference Series **2793** (2024) 011001 doi:10.1088/1742-6596/2793/1/011001

Organizing Committees

Conference Chairs
Naoyuki Ishimura, Chuo University, Japan
Po-Liang Liu, National Chung-Hsing University, Taiwan

Program Chairs
Kuo-Chih Chuang, Zhejiang University, China
Jen Chuan Tung, Chang Gung University, Taiwan
Girija Chetty, University of Canberra, Australia

TPC Members
Chukiat Chaiboonsri, Chiang Mai University, Thailand
Rene Batac, De La Salle University, Philippines
Ibrahim Rose Irnawaty, Universiti Sains Islam Malaysia, Malaysia
Alexandra B. Santos-Putungan, University of the Philippines Los Banos, Philippines
Mazhar Hussain, Government College University, Pakistan
Maria Cecilia Galvez, De La Salle University, Philippines
Natita Wangsoh, Pathumwan Institute of Technology, Thailand
Darwin Barayang Putungan, University of the Philippines Los Banos, Philippines
Thana Yeeram, Mahasarakham University, Thailand
Wachirapond Permpoonsinsup, Pathumwan Institute of Technology, Thailand
Xiaodi Li, Shandong Normal University, China
Agus Maman Abadi, Yogyakarta State University, Indonesia
Zailan Siri, University of Malaya, Malaysia
Nik Mohd Asri Nik Long, Universiti Putra Malaysia, Malaysia
Zayid Abdulhadi, American University of Sharjah, United Arab Emirates
Dalal Maturi, king Abdulaziz University, Saudi Arabia
Kwangwon Ahn, Yonsei University, South Korea
Rubén Darío Santiago Acosta, Tecnológico de Monterrey, Mexico
Bushra Aliwi, University of Babylon, Iraq
Khushboo Verma, University of Lucknow, India
Reza Mohammadi, University of Neyshabur, Iran
Joginder Dhiman, Himachal Pradesh University, India
Chaiya Luengviriya, Kasetsart University, Thailand
Madhusudhan Mishra, NERIST, India
Zety Sharizat Hamidi, Universiti Teknologi MARA, Malaysia
Johan Matheus Tuwankotta, Bandung Institute of Technology, Indonesia.
Mitali Konwar, Moran College, India
Thunyawadee Sucharidtham, Rajamangala University of Technology, Thailand
Tau Han Cheong, Universiti Teknologi MARA Malaysia, Malaysia
Cheng-Kang Le, Cheng Shiu University, Taiwan
Ahmed Elsaid, Egypt-Japan University of Science and Technology, Egypt
Burhan Tiryakioğlu, Marmara University, Turkey

doi:10.1088/1742-6596/2793/1/011001

Yu-Fen Chang, UiT the Arctic University of Norway, Norway
Nur Nafhatun Mohd Shariff, Universiti Teknologi MARA Shah Alam, Malaysia
Khreshna Syuhada, Institut Teknologi Bandung, Indonesia
Alina Minitaeva, Bauman Moscow State Technical University, Russia
Fuaada Mohd Siam, Universiti Teknologi Malaysia, Malaysia
Jonah Lissner, Technion - Israel Institute of Technology, Israel

Journal of Physics: Conference Series **2793** (2024) 011001 doi:10.1088/1742-6596/2793/1/011001

STATEMENT OF PEER REVIEW

All papers published in this volume of Journal of Physics: Conference Series have been peer reviewed through processes administered by the proceedings Editors. Reviews were conducted by expert referees to the professional and scientific standards expected of a proceedings journal published by IOP Publishing.

Journal of Physics: Conference Series **2793** (2024) 011002 doi:10.1088/1742-6596/2793/1/011002

Peer Review Statement

All papers published in this volume have been reviewed through processes administered by the Editors. Reviews were conducted by expert referees to the professional and scientific standards expected of a proceedings journal published by IOP Publishing.

- **Type of peer review:** Double Anonymous
- **Conference submission management system:** Morressier
- **Number of submissions received:** 50
- **Number of submissions sent for review:** 50
- **Number of submissions accepted:** 27
- **Acceptance Rate (Submissions Accepted / Submissions Received × 100):** 54
- **Average number of reviews per paper:** 1
- **Total number of reviewers involved:** 38
- **Contact person for queries:**
  Name: Ching SHEU
  Email: sheu.ching0704@gmail.com
  Affiliation: Asia University, Taiwan

Journal of Physics: Conference Series **2793** (2024) 012001 doi:10.1088/1742-6596/2793/1/012001

Numerical analysis clarifying non-equilibrated periodic flows generated by dual-scaling Taylor's analytical solution

Xuanyou Gong[1], Hiroki Suzuki[1,*] and Toshinori Kouchi[1]

[1]Graduate School of Environmental, Life, Natural Science and Technology, Okayama University, 3-1-1 Tsushima-naka, Kita-ku, Okayama-shi, Okayama 700-8530, Japan

* Corresponding author's email: h.suzuki@okayama-u.ac.jp

Abstract. This study investigates the properties of a flow field derived from the multiplexed Taylor solution, focusing on its impact on established equilibrium relationships. We numerically analyse a multiplexed flow field by introducing a secondary flow with twice the period of the original Taylor solution, hypothesising that this dualisation affects the balance between transient and viscous terms. The visualisation of the flow field and the time evolution of the kinetic energy are investigated. Using second and sixth order central differences and the fractional step method with BiCGStab, we analyse three flow fields with different coefficients in a two-dimensional periodic domain. The study shows that the dualisation of the flow field disturbs the equilibrium between the unsteady and viscous terms. The time-averaged kinetic energy shows that the evolution of the dualised field approaches the kinetic energy distribution of the analytical solution, with variations depending on the coefficient of the quadratic term. The spatial distribution of the kinetic energy in fully developed fields is also compared with the analytical solution.

1. Introduction

Understanding incompressible flows, which are prevalent in our environment, is crucial for modelling turbulent phenomena [1]. These turbulent fields often exhibit universal high-frequency fluctuations that are also subject to viscous dissipation. A key feature of these fluctuations is the variety of spatial scales involved; they are not confined to a single wavenumber, but typically involve multiple wavenumbers. Conversely, single wavenumber flow fields are simpler and more straightforward to analyse. This study therefore aims to elucidate the effects of multiplexing on these simpler single wavenumber flows.

A prime example of a single wavenumber flow field is the analytical Taylor solution, which describes a two-dimensional periodic flow and obeys the governing equations. This solution also allows the decay of kinetic energy due to viscous dissipation to be expressed by a simple exponential function. Its complete formulation makes the Taylor solution an excellent tool for validating numerical analyses, such as studying the effects of errors in the conservation laws [2]. The finite volume method, often coupled with discretisation techniques, is typically used to solve incompressible flows with high accuracy in the conservation laws. In wind tunnel experiments, turbulence is generated using a one-scale turbulence grid consisting of grid bars spaced at a constant distance [3-5], which defines the characteristic length of the bulk flow.

In this context, a fractal turbulence grid, created by multiplexing the conventional grid, is used to generate turbulence in wind tunnels [6-8]. This fractal grid often includes a square grid element in addition to the standard layout and has different scales, ranging from half to a quarter of the maximum scale. Our study applies this fractal concept to Taylor's analytical solutions. We construct a flow field

Journal of Physics: Conference Series **2793** (2024) 012001 doi:10.1088/1742-6596/2793/1/012001

by multiplexing the original Taylor solution and analyse this modified field. Previous research using the Taylor solution has shown that its governing equations involve two independent equilibrium relationships [2], where the unsteady term is balanced by the viscous term and the nonlinear term by the pressure term. Our study aims to clarify how multiplexing affects these equilibrium relationships.

The aim is to detail the properties of a flow field derived from the multiplexed Taylor solution, in particular its effect on the established equilibrium relationships. We apply numerical analysis to a multiplexed flow field by adding a secondary flow with twice the period of the original Taylor solution. This study hypothesises that such dualisation may affect the balance between transient and viscous terms. We examine the visualisation of the flow field and the time evolution of the kinetic energy. Since the kinetic energy in the Taylor solution can be represented by a simple exponential function, the study also examines how the doubling of the flow field affects the distribution of the kinetic energy.

2. Numerical methods

This study focuses on the analytical Taylor solution. The Taylor analytical solution has been used in previous studies. Components of a velocity field, u and v for the analytical Taylor solution is given by

$$u(x,y,t) = -2\cos(x)\sin(y)\exp(-2t/\mathrm{Re}) \text{ and } v(x,y,t) = 2\sin(x)\cos(y)\exp(-2t/\mathrm{Re}) \quad (1)$$

Here, x, y, t are spatial and temporal directions, respecitvely, and Re is the Reynolds number. As shown in the above formula, the spatial distribution of the analytical Taylor solution is composed of sinusoidal functions and the time evolution is given by an exponential function. Using the velocity field in the above formula, the kinetic energy per unit area $K(t)$ is given by

$$K(t) = (1/2)(\langle u^2\rangle + \langle v^2\rangle) = \exp(-4t/\mathrm{Re}), \text{ which is for Case 0.} \quad (2)$$

Here $\langle\ \rangle$ is the ensemble mean. In other words, the integral value per area of the kinetic energy at each position is obtained. As shown in the above equation, this kinetic energy is given by a simple exponential function as a function of time.

The analytical Taylor solution satisfies the Navier-Stokes equation and the continuity equation. The unsteady terms of Navier-Stokes are given below when the velocity field is that of the Taylor analytic solution.

$$\partial u/\partial t = (4/\mathrm{Re})\exp(-2t/\mathrm{Re})\cos(x)\sin(y) \text{ and } \partial v/\partial t = -(4/\mathrm{Re})\exp(-2t/\mathrm{Re})\sin(x)\cos(y) \quad (3)$$

On the other hand, the viscous term of the Navier-Stokes equations is given for the velocity field of the analytical Taylor solution as follows

$$\begin{aligned}(1/\mathrm{Re})(\partial^2 u/\partial x^2 + \partial^2 u/\partial y^2) &= (4/\mathrm{Re})\exp(-2t/\mathrm{Re})\cos(x)\sin(y)\\ (1/\mathrm{Re})(\partial^2 v/\partial x^2 + \partial^2 v/\partial y^2) &= -(4/\mathrm{Re})\exp(-2t/\mathrm{Re})\sin(x)\cos(y)\end{aligned} \quad (4)$$

As can be seen from the above two equations, the unsteady and viscous terms of the Navier-Stokes equation for the velocity field of the Taylor analytical solution are identical. This indicates that the unsteady and viscous terms are balanced in the flow field described by the analytical Taylor solution. Furthermore, this balance indicates that the non-linear and pressure terms are balanced when this analytical solution is obtained. Therefore, when the analytical Taylor solution is obtained, there are two independent equilibrium relationships in the flow field.

This study investigates the time evolution of this analytical Taylor solution in a flow field with a doubling flow field as the initial field. The initial field of the analytical Taylor solution is given by

$$u(x,y,t) = -2\cos(x)\sin(y) \text{ and } v(x,y,t) = 2\sin(x)\cos(y) \quad (5)$$

By doubling this flow field, the following flow field is obtained.

Journal of Physics: Conference Series **2793** (2024) 012001 doi:10.1088/1742-6596/2793/1/012001

$$u(x,y,t) = -\ C_a\,(\cos(x)\sin(y) - \cos(mx)\sin(my))$$
$$v(x,y,t) = \ C_a\,(\sin(x)\cos(y) + \sin(mx)\cos(my)) \quad (6)$$

where m is an integer larger than or equal to 2. The constant C_a is included in the above equation. The kinetic energy per unit area of the velocity field shown in the above equation is given by

$$K(t) = (C_a{}^2/2)\exp(-\,4t\,/\,\mathrm{Re}),\ \text{which is for Case 1a.} \quad (7)$$

By equating the value of this equation with that of the Taylor analysis solution, the value of C_a can be obtained as follows: $C_a = 2^{1/2}$. In addition to the one given above, this study considers that there are two other ways of giving the second scale flow field. Two additional doubled flow fields are given by this study as follows.

$$u(x,y,t) = -\ C_b\,(\cos(x)\sin(y) - (1/m)\cos(mx)\sin(my))$$
$$v(x,y,t) = \ C_b\,(\sin(x)\cos(y) + (1/m)\sin(mx)\cos(my)) \quad (8a)$$

$$u(x,y,t) = -\ C_c\,(\cos(x)\sin(y) - (1/m^2)\cos(mx)\sin(my))$$
$$v(x,y,t) = \ C_c\,(\sin(x)\cos(y) + (1/m^2)\sin(mx)\cos(my)) \quad (8b)$$

The difference between the above three flow fields lies in the way the integer coefficient m depends on the second term. The constants C_b and C_c in the above equations are given below by equating the kinetic energy of the doubled velocity field to that of the Taylor analysis solution.

$$K(t) = (C_b{}^2(m^2 + 1)/(4\,m^2))\exp(-\,4t\,/\,\mathrm{Re}),\ \text{which is for Case 1b.}$$
$$K(t) = (C_c{}^2(m^4 + 1)/(4\,m^4))\exp(-\,4t\,/\,\mathrm{Re}),\ \text{which is for Case 1c.} \quad (9)$$

The properties of the above three dualised velocity fields are summarised as Case 1a, 2b and 1c, which are set in the present numerical analysis.

For the dualised flow field thus given, the unsteady and viscous terms of the Navier-Stokes equations are obtained: for the dualised velocity field in Case 1a, the unsteady and viscous terms with an assumption on the temporal characteristics of exp(– 2 t/Re) for the velocity components are obtained as follows.

$$\partial u/\partial t = \ C_a\,(2\,/\mathrm{Re})\exp(-\,2t\,/\,\mathrm{Re})\,[\cos(mx)\sin(my) + \cos(mx)\sin(my)]$$
$$\partial v/\partial t = -\ C_a\,(2\,/\mathrm{Re})\exp(-\,2t\,/\,\mathrm{Re})\,[\sin(mx)\cos(my) + \sin(mx)\cos(my)] \quad (10a)$$

$$(1/\mathrm{Re})(\partial^2 u/\partial x^2 + \partial^2 u/\partial y^2) = \ C_a\,(2/\mathrm{Re})\exp(-\,2t/\mathrm{Re})\,[m^2\cos(mx)\sin(my) + \cos(mx)\sin(my)]$$
$$(1/\mathrm{Re})(\partial^2 v/\partial x^2 + \partial^2 v/\partial y^2) = -\ C_a\,(2/\mathrm{Re})\exp(-\,2t/\mathrm{Re})\,[m^2\sin(mx)\cos(my) + \sin(mx)\cos(my)] \quad (10b)$$

As shown in the above equation, the unsteady and viscous terms are not identical, as opposed to the case of the analytical Taylor solution. This indicates that the equilibrium relationship between the two terms established in the analytical Taylor solution is not established in the doubled flow field. At the same time, this also indicates that the equilibrium relationship between the non-linear term and the pressure term, established in the Taylor analytical solution, is also not established. This non-equilibration caused by the doubling of the velocity field is also observed in Case 1b and Case 1c. These results show that

Journal of Physics: Conference Series **2793** (2024) 012001 doi:10.1088/1742-6596/2793/1/012001

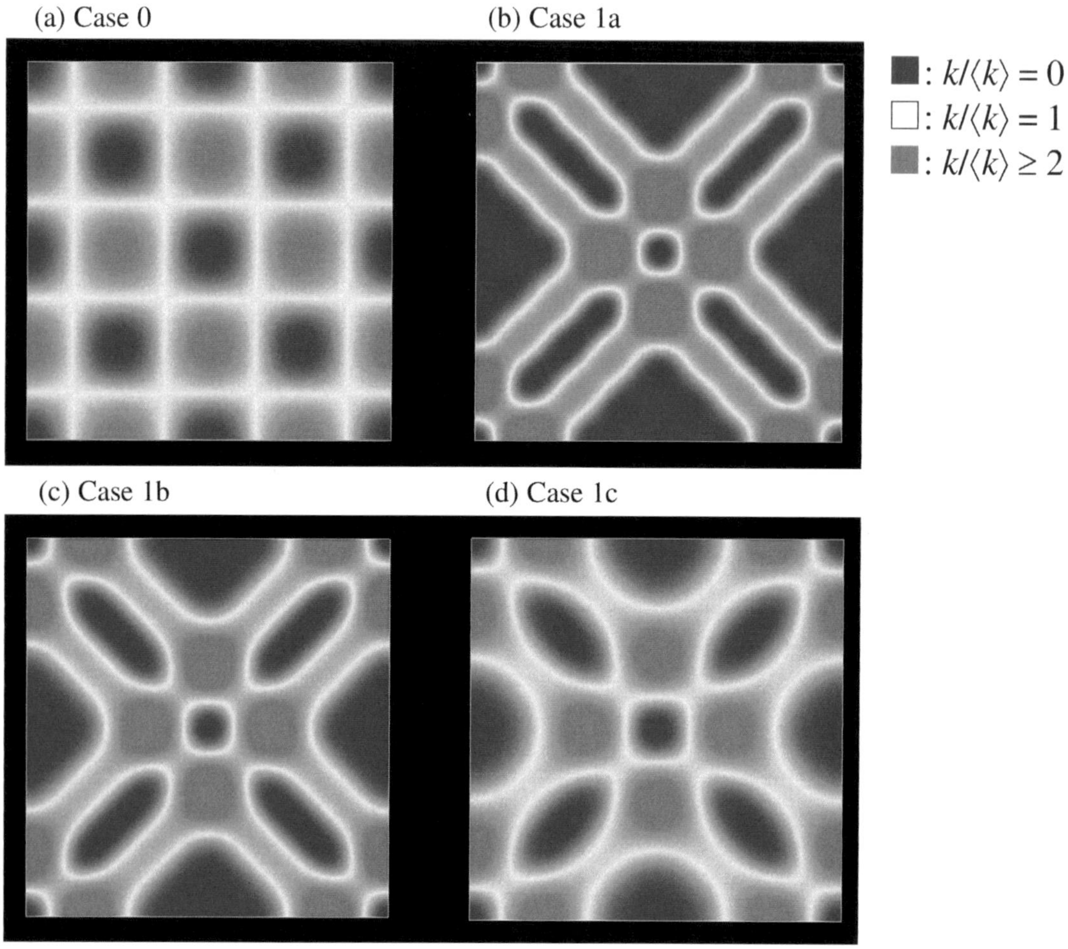

Figure 1. Initial kinetic energy fields for the four cases.

the doubling of the velocity field of the analytical Taylor solution leads to the non-equilibration of the flow field of the analytical Taylor solution.

The object of analysis is an incompressible flow in a two-dimensional periodic domain. Here one side of the domain is 2π. This analysis domain is set up to solve numerically for the Taylor analytic solution. The governing equations for this flow are the continuity equation and the Navier-Stokes equations. A fractional step method is used to solve these equations. The six-step fourth order accurate Runge-Kutta method is used to integrate the governing equations in time. The BiCGStab method is used for the poisson equation at a fractional step. The error in the continuity equation is of machine zero order. The convection and pressure terms in the governing equations are discretised using a second-order accuracy central difference method. This discretisation method is identical to that used in previous studies [2,9]. The kinetic energy conservation errors due to the convection and pressure terms are sufficiently small in the present study. The viscosity terms are discretised using a sixth order accuracy central difference method [10]. As a computational condition, an boundary of the computational domain is discretised with 128 grid points. The time step is 0.001 and the equations are integrated in time until the dimensionless time is 2π. The Reynolds number is 2π. In this study the parameter for the doubling of the flow field is set to $m = 2$.

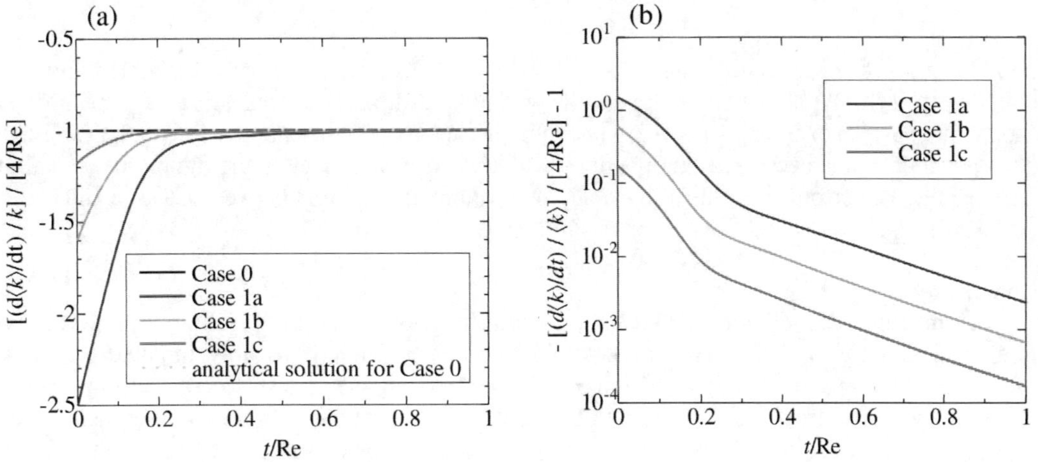

Figure 2. Time variations of a quantity to evaluate if the spatially averaged kinetic energy can be described by an exponential function, (a). Here, the exponent is - 4/Re in the analytical solution. (b) Time variations of relative deviations from the exponent in the analytical solution.

3. Results and Discussion

Figure 1 shows the visualisation of the kinetic energy map of the initial field of the analytical Taylor solution. As shown in the figure, lines represented by coincident values of the mean kinetic energy are arranged in a grid. The squares of the grid then show alternating regions of high and low energy. This kinetic energy distribution is significantly modified by doubling: as shown in the results of Case 1a, the kinetic energy of the doubled flow field shows a distribution similar to two diagonal lines alternately connected, where the grid distribution is seen. In contrast to the analytical Taylor solution, case 1a shows a distribution of the flow field based on multiple scales.

Figure 1 also shows the kinetic energy map of the dualised initial flow fields for Case 1b and Case 1c. As shown in the figure, in Case 1b and Case 1c, as in Case 1a, the grid-like distribution seen for the Taylor analysis solution is not observed, but a distribution along two diagonals. In Case 1c, unlike in Cases 1a and 1b, there is no kinetic energy distribution parallel to the two diagonals. As shown above, the present study shows that the doubling forms a flow field with a distribution far from that of the Taylor analysis solution, and that the distribution varies depending on the multiplexing exponent *m* applied.

In order to rigorously test whether the kinetic energy distribution due to the doubled initial field can be described by a single exponential function, the present study considers the following quantity: $[d\langle k\rangle/dt/k]$. This quantity is strictly constant with time if the distribution is exponential. Furthermore, in the flow field of the analytical solution, this quantity coincides with - 4/Re. In Figure 2, the value is divided by 4/Re, so the result of the analytical solution is equal to -1. In Case 0, where the initial field of the analytical Taylor solution is set, the value is in good agreement with -1, the value corresponding to the analytical solution; in Cases 1a to 1c, the value changes with time at the initial time. This indicates that the kinetic energy distribution at this stage of development is not described by an exponential function. Here the deviation of the values from the analytical solution differs between Case 1a and Case 1c.

After the flow field has developed, in Cases 1a to 1c, the value obtained for this quantity is constant with time. In addition, the value of this quantity is sufficiently close to the value of the analytical solution. These results show that in a sufficiently developed time range, the kinetic energy distribution of the multiplexed flow field asymptotically approaches an exponential function with the exponent of the analytical solution, as shown in the previous figure. The deviation of the exponent of the exponential function from that of the analytical solution is shown as a function of time when a doubled initial field is set up in Fig. 2 (b). As shown in the figure, the deviation of the exponential values is large during the development phase up to about $t/\mathrm{Re} = 0.2$. Thereafter, at times larger than about $t/\mathrm{Re} = 0.2$, the deviation of the exponential value decreases exponentially with time, as can be seen in the figure: the deviation of the exponential value from the value of the analytical solution is different between Case 1a and Case 1c, however the shape of the time distribution is similar.

4. Conclusion

The aim of this study was to understand the non-equilibrium flow properties resulting from the doubling of the flow field of the analytical Taylor solution. The time evolution of the flow, initiated by dualising the analytical solution and using it as the initial velocity field, showed a disturbed equilibrium between unsteady and viscous terms. Three flow fields with different coefficients were analysed using second and sixth order central differences and the fractional step method with BiCGStab for the Poisson equation. A two-dimensional periodic domain based on the Taylor solution was used. The research visualised the kinetic energy distribution in the initial field, revealing differences from the analytical field. The time-averaged kinetic energy showed that the development stage of the doubled field and its asymptotic approach to the kinetic energy distribution of the analytical solution. The deviation of the kinetic energy distribution from the analytical solution varied with the coefficient of the quadratic term, but their asymptotes were similar. The study also compared the spatial kinetic energy distribution in fully developed fields from the doubled flow with the analytical solution.

Acknowledgments

The author is deeply grateful to Dr Kento Tanaka (Okayama University) for his support in carrying out this work. This work was partly supported by the Japanese Ministry of Education, Culture, Sports, Science and Technology through Grants-in-Aid (Nos. 21K03859 and 22H01684).

References

[1] Pope S B 2000 Turbulent Flows. Cambridge University Press

[2] Suzuki H, Hattori S, Mochizuki S 2017 Numerical investigation using an exact solution of the effects of non-solenoidality of the viscous terms on the incompressible flow *J. Fluid Sci. Tech.* 12 (1) 16-00257

[3] Suzuki H, Mochizuki S, Hasegawa Y 2016 Validation scheme for small effect of wind tunnel blockage on decaying grid-generated turbulence *J. Fluid Sci. Tech.* 11 (3) 16-00195

[4] Mohamed M S, Larue J C 1990 The decay power law in grid-generated turbulence *J. Fluid Mech.* 219 195-214

[5] Kitamura T, Nagata K, Sakai Y, Sasoh A, Terashima O, Saito H, Harasaki T 2014 On invariants in grid turbulence at moderate Reynolds numbers *J. Fluid Mech.* 738 378-406

[6] Nagata K, Sakai Y, Inaba I, Suzuki H, Terashima O, Suzuki H 2013 Turbulence structure and turbulence kinetic energy transport in multiscale/fractal-generated turbulence *Phys. Fluids* 25 (6) 065102

[7] Valente P C, Vassilicos J C 2011 The decay of turbulence generated by a class of multiscale grids *J. Fluid Mech.* 687 300-340

[8] Hearst R J, Lavoie P 2014 Decay of turbulence generated by a square-fractal-element grid *J. Fluid Mech.* 741 567-584

[9] Morinishi Y, Lund T S, Vasilyev O V, Moin P 1998 Fully conservative higher order finite difference schemes for incompressible flow *J. Comput. Phys.* 143(1) 90-124

[10] Berland J, Bogey C, Bailly C 2006 Low-dissipation and low-dispersion fourth-order Runge–Kutta algorithm *Comput. Fluids* 35(10) 1459-1463

Journal of Physics: Conference Series **2793** (2024) 012002 doi:10.1088/1742-6596/2793/1/012002

Can Newtonian kinetic energy and Einsteinian rest-mass energy be expressed by the binomial expansion of the Lorentz factor? And how valid is using Einstein's $E = mc^2$ to calculate the nuclear fission energy?

Talal Al-Ameen[1*], Maythem Mahmud[2] and Imad Muhi[3]

[1,2,3] Department of Mathematics and Natural Sciences, Prince Mohammad Bin Fahd University (PMU), Al Khobar, KSA

* Corresponding author's email: tameen@pmu.edu.sa

Abstract. The binomial (Taylor) expansion of the Lorentz factor has been reconsidered here in an attempt to find out whether the *Newtonian kinetic energy* and the *Einsteinian rest-mass energy* are implicitly embedded in the mathematical structure of the binomial expansion of the Lorentz factor (as Einstein postulated in his Special Theory of Relativity). Advocates of Standard Special Relativity show that it is possible to obtain these two kinds of energy by multiplying both sides of the expansion of the Lorentz factor by the moving object's rest mass m_0 and the square of the speed of light c^2. This study shows that the *apparent* reconciliation between classical and relativistic physics made possible by employing the binomial expansion of the Lorentz factor $\gamma = [1 - (\frac{v^2}{c^2})]^{-\frac{1}{2}}$, where v is the moving object's velocity, is challengeable. Also it is unclear how Einstein's famous rest-mass energy equation $E = m_0c^2$ from $E_{kinetic\ net} = m_{relativistic}c^2 - m_0c^2$ can be used to calculate the amount of nuclear fission energy released?

1. Introduction

Standard Relativity claims that Newtonian equations can be obtained from relativistic ones. In particular it has been assumed that relativistic kinetic energy may be approximated to classical kinetic energy within the limits of low velocities ($\frac{v}{c} < 1$, where v is the velocity of a moving body, and c the velocity of light). This *apparent* reconciliation between classical and relativistic physics was made possible by employing the binomial (and/or Taylor) expansion of the Lorentz factor $\gamma = [1 - (\frac{v^2}{c^2})]^{-\frac{1}{2}}$. This factor was postulated by both FitzGerald (in 1889) and Lorentz (in 1892) and appears in all mathematical works that deal with kinematics of relative motions, stating that objects (bodies and particles) in the direction of their motion contract exactly in accordance with the rate of Lorentz Factor [1,2]. In 1905, Einstein derived the Lorentz transformation equations[3].

This paper is organized as follows. In section 2, the problem statement of the theoretical work of the paper is outlined. In Sections 3, Lorentz's definition of simultaneity and an alternative derivation to his original derivation (of the factor) are reproduced. Section 4 presents how valid is using Taylor expansion to show that relativistic kinetic energy may be approximated to classical kinetic energy within the limits of low velocities ($\frac{v}{c} < 1$)? Also this section argues the validity of using Einstein's $E = m_0c^2$ (where m_0 is the object's rest mass, and c the square of the speed of light) to calculate the amount of fission energy released. Finally, section 5 presents some concluding remarks, followed by references relative to the problems considered.

2. Statement of the problems

In the Lorentz factor, $\gamma = [1 - (\frac{v^2}{c^2})]^{-\frac{1}{2}}$, mathematical singularities ($\frac{1}{0}$) become unavoidable if $\frac{v^2}{c^2} \to 1$. Therefore, it is more appropriate to express it by the binomial expansion (approximation) for $\frac{v}{c} < 1$ as $\gamma = [1 - (\frac{v^2}{c^2})]^{-\frac{1}{2}} = 1 + \frac{v^2}{2c^2} + \frac{3\,c^4}{8\,v^4} + \ldots$ where the $(\ldots)$ (dots) stand for smaller corrections we are ignoring.

Can this expansion reconcile classical and relativistic physics as far as Newtonian kinetic energy and Einsteinian rest-mass energy are concerned? And how valid is using Einstein's $E = m_0c^2$ to calculate the amount of fission energy released?

3. The Mathematics of Lorentz factor

If you, sitting at (x', t'), passed by me, sitting at (x, t), by a velocity v when $t' = t$, see Figure 1, below, then the relationship between your x' and my x is:

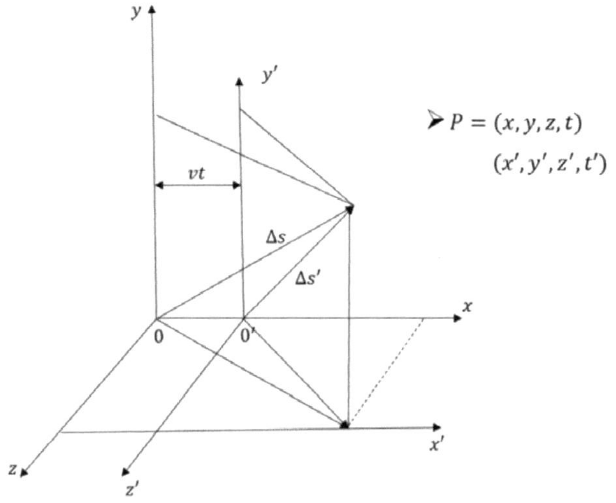

Figure 1. For any event, say P, the space-time coordinates of P in my frame of reference and your frame of reference are $(x, y. z, t)$ and (x', y', z', t') respectively.

$$x' = x + vt \quad \text{(your position is at distance } vt \text{ away from mine).}$$

$$x = x' - vt \quad \text{(my position is at distance } vt \text{ away from yours).}$$

For the velocity of light c to be same for both of us, then for you:

$$x' = \gamma(x + vt) = \gamma(ct + vt) = ct' \quad \text{where } x = ct$$

And for me:

$$x = \gamma(x' - vt) = \gamma(ct' - vt) = ct \quad \text{where } x' = ct'.$$

$(ct)\,(ct') = c^2 t\, t' = \gamma^2(ct' - vt)\,(ct + vt) = \gamma^2(ct' - vt')\,(ct + vt)$ (Substituting $vt' = vt$, as already given $t' = t$).

$$c^2 t\, t' = \gamma^2 t\, t'(c - v)\,(c + v) = \gamma^2 t\, t'(c^2 - v^2)$$

$$c^2 = \gamma^2(c^2 - v^2) = \gamma^2(c + v)(c - v)$$

$$\gamma = \pm[1 - (\frac{v^2}{c^2})]^{-\frac{1}{2}} = 1 + \frac{v^2}{2c^2} + \frac{3v^4}{8c^4} + \ldots$$

Note that the term $(c^2 - v^2)$ in γ^2 is $(c + v)(c - v)$**.** No matter how v is small compared to c, the terms $(c + v)$ and $(c - v)$ suggest that we can *add* to c or *subtract* from it. How possible is this? The answer as per Special Relativity is "the speed of light is independent of the relative motion of source and observer, whether the observer is on a rocket (cruising with $v \leq c$) or Earth-bound. No velocities can be added to greater than the speed of light, provided that v is less than c."

Mathematically, what is $(c + v)$ if not a velocity that is greater than c if $v > 0$? And what is $(c - v)$ if not a velocity that is smaller than c if $v > 0$? This realization went without have been noticed or discussed in the literature.

4. Newtonian kinetic energy and Einsteinian rest-mass energy

As per the binomial expansion, $\gamma = [1 - (\frac{v^2}{c^2})]^{-\frac{1}{2}} = 1 + \frac{v^2}{2c^2} + \frac{3v^4}{8c^4} + \ldots$ where $\frac{v}{c} < 1$. Multiplying both sides of this relation by $m_0 v$, rather than $m_0 c$, because mathematically the expansion holds only, and only, for $v < c$, and ignoring the inconsequential terms such as $\frac{3v^4}{8c^4} + \ldots$ as they are relatively small compared to $(1 + \frac{v^2}{2c^2})$, we obtain:

$$\gamma m_0 v = m_0 v + \frac{m_0 v^3}{2c^2} + \ldots$$

Theoretically, the only acceptable way to change this relativistic momentum relation to relativistic energy relation is to multiply it by v (the velocity of the moving mass-ive object) and not the speed of light c (because the binomial expansion holds only for $v < c$).

$$\gamma m_0 v^2 = m_0 v^2 + \frac{m_0 v^4}{2c^2} + \ldots$$

Thus, the assumption in Standard Special Relativity that when $v = 0$ there is a rest-mass energy $(m_0 c^2)$ cannot be supported.

With this correction it is evident that Standard Special Relativity must accept the conclusions of Classical Newtonian kinematics in the view that when $v = 0$ there can be no rest energy $(E_0 = m_0 c^2)$. It is not our intention to argue the accuracy of this view but only to demonstrate that if Standard Special Relativity adopts the above form of Taylor expansion then it must base its conclusions upon correct mathematical dealings.

We may ask the following two questions to further assess the validity of incorporating $\gamma = [1 - (\frac{v^2}{c^2})]^{-\frac{1}{2}} \approx 1 + \frac{v^2}{2c^2}$ in Standard Special Relativity:

I) Can the above argument truly invalidate the existence of rest energy (m_0c^2) when $v = 0$?

Advocates of Standard Special Relativity would respond "*not really*!" This is because, as per their mathematics, the outcome of multiplying both sides of $\gamma = [1 - (\frac{v^2}{c^2})]^{-\frac{1}{2}} \approx 1 + \frac{v^2}{2c^2}$ by m_0 and c^2 respectively would give:

Step-1 (*the relativistic mass*): $\gamma m_0 = m_0[1 - (\frac{v^2}{c^2})]^{-\frac{1}{2}} = m_0 + \frac{m_0v^2}{2c^2}$ +...

Step-2 (*the relativistic energy*): $\gamma m_0c^2 = m_0c^2[1 - (\frac{v^2}{c^2})]^{-\frac{1}{2}} = m_0c^2 + \frac{m_0v^2}{2}$ +...

where m_0c^2 is the Einstein rest-mass energy, and $\frac{m_0v^2}{2}$ the Newtonian kinetic energy.

But if the multiplication $\gamma m_0 = m_0[1 - (\frac{v^2}{c^2})]^{-\frac{1}{2}} = m_0 + \frac{m_0v^2}{2c^2}$ +... is allowed, and knowing that the Lorentz factor is a term by which relativistic mass, time, and length changes for an object in motion as measured in two different frames of reference, then by the same token it would be perfectly legitimate to multiply $\gamma = [1 - (\frac{v^2}{c^2})]^{-\frac{1}{2}} \approx 1 + \frac{v^2}{2c^2}$ by the moving object's *cubic rest length* (L_0^3) in the direction of the motion [bearing in mind that L_0 is as fundamental physical quantity as m_0 and goes without saying that the rest mass m_0 is entirely (for a 3- dimensional object) confined in L_0^3], giving:

$$\gamma L_0^3 = L_0^3 + \frac{L_0^3v^2}{2c^2} + ...$$

But, how valid is this assumption theoretically?

For a sufficiently very rigid spherical object of radius r_0 at rest, L_0 can be said equal to ($L_0 = r_0$), and the rest mass of the sphere can be expressed as $m_0 = \int 4\pi {r_0}^2 \rho \, dr_0$, where ρ is the object's density, resulting (after integration) in:

$$m_0 \propto r_0^3 \text{ (where } \propto \text{= proportional to)}$$

$$\gamma m_0 \propto \gamma r_0^3 \text{ (multiplying both sides by } \gamma)$$

$$m_0[1 - (\tfrac{v^2}{c^2})]^{-\frac{1}{2}} \propto r_0^3[1 - (\tfrac{v^2}{c^2})]^{-\frac{1}{2}}$$

Or

$$m_0 + \frac{m_0v^2}{2c^2} + ... \propto r_0^3 + \frac{r_0^3v^2}{2c^2} + ...$$

The implication of this, for Lorentz-Einstein's equations, would seriously challenge their established proof for *length contraction*, $r = r_0[1 - (\frac{v^2}{c^2})]^{\frac{1}{2}}$, in the Standard Special Relativity. Rather than confirming *length contraction* ($r \to 0$ when $v \to c$), the expansion [$r_0^3 + \frac{r_0^3v^2}{2c^2} + ...$] demands the opposite i.e. *length elongation* (when $v \to c$)!

How can a zero radius sphere [$r \to 0$ when $v \to c$ in $r = r_0[1 - (\frac{v^2}{c^2})]^{\frac{1}{2}}$] have apparently infinite mass [$m_{relativistic} \to \infty$ when $v \to c$ in $\gamma m_0 = m_0[1 - (\frac{v^2}{c^2})]^{-\frac{1}{2}}$ if $m_0 > 0$] ?

When the length of one dimensional moving object tends to zero (in the direction of motion), the object disappears from existence (in that direction), and we cannot say that its mass tends to infinity. An infinite mass demands an infinite space (length). If so, the proportionality relation of $m_0 + \frac{m_0 v^2}{2c^2} +.. \propto r_0^3 + \frac{r_0^3 v^2}{2c^2} +...$ is more realistic and logical than $\gamma m_0 = m_0[1-(\frac{v^2}{c^2})]^{-\frac{1}{2}}$ for $m_0 > 0$ with respect to $\frac{r}{\gamma} = r_0[1-(\frac{v^2}{c^2})]^{\frac{1}{2}}$ for $r_0 > 0$.

It seems more realistic to say that the object's relativistic kinetic energy can increase with increasing speed, but surely not the object's rest mass. If so, then the validity of $\gamma m_0 = m_0[1-(\frac{v^2}{c^2})]^{-\frac{1}{2}}$ for $m_0 > 0$ is uncertain and questionable.

Was Einstein himself at ease with the concept of relativistic mass γm_0 which he introduced into Newton physics?

The letter below written by Einstein demonstrates his own *disquiet* with the concept of relativistic mass $\gamma m_0 = m_0[1-(\frac{v^2}{c^2})]^{-\frac{1}{2}} = m_0 + \frac{m_0 v^2}{2c^2} + ...$ and thereby with the relativistic energy $\gamma m_0 c^2 = m_0 c^2[1-(\frac{v^2}{c^2})]^{-\frac{1}{2}} = m_0 c^2 + \frac{m_0 v^2}{2} +...$ too in his Special Relativity!

"It is not good to introduce the concept of the (relativistic) mass $M = m_0[1-(\frac{v^2}{c^2})]^{-\frac{1}{2}}$ of a body for which no clear definition can be given. It is better to introduce no other mass than 'the rest mass' m. Instead of introducing M, it is better to mention the expression for the momentum and energy of the body in motion." (Albert Einstein's letter to Lincoln Barnett in 1948).

II) Can the expansion $\gamma = [1-(\frac{v^2}{c^2})]^{-\frac{1}{2}} = 1 + \frac{v^2}{2c^2} +...$ prove the existence of Einsteinian rest-mass energy ($m_0 c^2$) when $v = 0$ and the Newtonian kinetic energy ($E_k = \frac{m_0 v^2}{2}$) when $v > 0$?

To validate Step-2 (see above), the advocates of Standard Special relativity claim that the relativistic expression for kinetic energy E_k can be obtained from the work-energy theorem. Specifically, if a force expressed as $F = \frac{m_0 d(\gamma v)}{dt}$ moves a body from rest ($v = 0$) to its final velocity ($v = v$), the work done on the body should be equal to its final kinetic energy. In mathematical form, for one-dimensional (x-axis) motion:

$$E_{kinetic} = \int F dx \quad \text{where } F = \frac{m_0 d(\gamma v)}{dt}$$

$$E_{kinetic} = \int m_0[\frac{d(\gamma v)}{dt}]\, dx = \int m_0 \left[\frac{d(\gamma v)}{dt}\right]\left(\frac{dx}{dt}\right) dt = \int m_0 \left[\frac{d(\gamma v)}{dt}\right] v\, dt$$

$$E_{kinetic} = m_0 \int v[\frac{d(\gamma v)}{dt}]dt$$

Integrating this equation by parts (from $v = 0$ to $v = v$) gives $E_{kinetic} = (\gamma - 1)m_0 c^2$

Substituting $\gamma \approx 1 + \frac{v^2}{2c^2}$ or $\gamma - 1 \approx \frac{v^2}{2c^2}$ in $E_{kinetic} = (\gamma - 1)m_0 c^2$ gives $E_{kinetic} = \frac{m_0 v^2}{2}$ ($v > 0$).

Now, how consistent is using the relation $F = \frac{m_0 d(\gamma v)}{dt}$ to arrive at $E_{kinetic} = \frac{m_0 v^2}{2}$?

Inconsistent! We believe.

The reason for this is that the relation $F = \frac{m_0 d(\gamma v)}{dt}$ involves *acceleration* $(\frac{dv}{dt})$. This does not mean that Standard Special Relativity, as the theory of flat Minkowski spacetime, is invalid in the presence of acceleration. But, in this theory, maintaining zero acceleration (constant linear velocity) in working out the *time dilation* and *length contraction* (of the moving object) while allowing non-zero acceleration $(\frac{dv}{dt} \neq 0)$ in working out $E_{kinetic} = \frac{m_0 v^2}{2}$ $(v > 0)$ (of the same object moving in the same frame of reference) is by all means a major inconsistency (was the object moving at constant velocity or accelerating (it cannot be both at the same time)?. Furthermore, the velocity v in $\gamma = [1 - (\frac{v^2}{c^2})]^{-\frac{1}{2}}$ is not momentous but rather a constant velocity (bearing in mind, Lorentz transformation is the only possible transformation between two inertial coordinate systems moving with *constant velocity* with respect to each other). It is the relative velocity between frames of reference (object's velocity is the same as its moving frame) in the direction of motion (x-direction). Or, alternatively, it is a parameter of transformation, and for a given boost it is a constant number, but can take a continuous range of (constant) values. We should not forget that the *Lorentz transformations* are a six-parameter family of linear transformations from a coordinate frame in spacetime to another frame that moves at a *constant velocity* relative to the former. Furthermore, the relation $F = \frac{m_0 d(\gamma v)}{dt} = \frac{d(\gamma m_0 v)}{dt}$ has the relativistic mas γm_0 which Einstein himself was not at ease with.

The relation $E_{kinetic\,net} = (\gamma - 1)\, m_0 c^2 = \gamma m_0 c^2 - m_0 c^2 = m_{relativistic} c^2 - m_0 c^2$ represents $W_{net} = E_{kinetic\,final} - E_{kinetic\,initial}$, where W_{net} (or $E_{kinetic\,net}$) is the net work done (or the net kinetic energy); $E_{kinetic\,final}$ (or $m_{relativistic}\ c^2$) is the (relativistic) final kinetic energy; and $E_{kinetic\,initial}$ (or $m_0 c^2$) is the initial state kinetic energy. Consistency, however, demands that kinetic energies (not atomic binding energy or rest-mass energy) should be subtracted from each other to have a net kinetic energy). If so, then Einstein's concept of $(m_{relativistic}\ c^2 - m_0 c^2)$ is an arbitrary definition and does not blend with $E_{kinetic\,net} = \frac{1}{2}\ m_0 v^2$! Unfortunately, this postulation has entered into convention without ever having been scrutinized or even had adequate discussion in theoretical physics. It has led to the widely accepted, yet confusing, statement that "mass is energy, and energy is mass: $E = mc^2$." The clear objection to this statement is the *photon.* Photon has energy, but it does not have mass. If so, then this statement must read: "all objects that have mass may have energy, but the converse may not necessarily be true."

Interestingly, nuclear fission (the source of the nuclear energy) is *non-relativistic* (i.e. $v << c$, or better still $v \approx 0$). If so, $E = mc^2$ cannot (in any mathematical way) be related directly to the creation of the nuclear bomb through fission.

In physics, $E = mc^2$ is used arbitrarily to calculate the amount of energy released from the nuclear bomb. The bomb equation is:

$$\text{neutron} + U^{235} \to U^{236} \to Kr^{91} + Ba^{142} + E + 3 \text{ neutrons out}$$

$$E = [\text{rest mass of } U^{236} - (\text{rest mass of } Kr^{91} + \text{rest mass of } Ba^{142})]\, c^2$$

Mathematically, it is unclear how $E = m_0 c^2$ from $E_{kinetic\,net} = m_{relativistic}\ c^2 - m_0 c^2$ can be called the amount of the nuclear fission energy that can be released from the atom? Rest mass of a stable atom is always less than the sum of rest masses of its nucleons. The difference is called mass defect Δm where the energy $\Delta E_0 = \Delta m_0 c^2$ is utilized in keeping the nucleons bound together. This energy, known as binding energy, is completely different from $E = m_0 c^2$.

The equation $E = mc^2$ (and we should not forget that it has physically nothing to do with nuclear fission and even nuclear energy release through radioactivity) says mathematically beyond any subjective interpretations that "$m = 1$ Kg of U^{236}" has the same amount of energy as "$m = 1$ Kg of anything else (e.g. cabbage)!" If we accept such result, then it leads to some rather *absurd* conclusions such as: 1 Kg U^{236} and 1 Kg of cabbage both have equal amount of energy:

$$E_{(m=1\text{ Kg})} = c^2 \approx 10^{17} \text{ Joule of energy, equivalent of 21 Megatons of TNT.}$$

To escape this conundrum, the defenders of $E = mc^2$ say subjectively, "the $E = mc^2$ is the internal nuclear binding energy of the atom''. We categorically believe it is not. No math and/or definition as we have already shown and discussed here can support this.

5. Conclusions

If Lorenz factor involves implicitly adding a speed to, or subtracting a speed from, the speed of light $[(c^2 - v^2)= (c + v)(c - v)]$, then this would mean that this factor violates the principle of Relativity, which does not allow adding a speed to, or subtracting a speed from, the speed of light. If this is true, then this factor is unreliable to be used in a theory of Relativity.

The length contraction of a moving body in the direction of motion would lead to the possibility of the disappearance of the length eventually when the velocity of the body approaches the speed of light. The disappearance of the length of a body from existence contradicts the increase in the mass of the body in that direction. An infinite mass cannot be confined in a dimension (length) that does not exist. The problem, we believe, lies in the derivation of the relativistic mass. Even Albert himself did not like the implication of relativistic mass (as it is evident from his letter to Lincoln Barnett in 1948).

Mathematically, the rest-mass energy m_0c^2 in $\gamma m_0c^2 = m_0c^2[1 - (\frac{v^2}{c^2})]^{-\frac{1}{2}} = m_0c^2 + \frac{m_0v^2}{2} + \dots$ initiates from the relativistic mass $\gamma m_0 = m_0[1 - (\frac{v^2}{c^2})]^{-\frac{1}{2}} = m_0 + \frac{m_0v^2}{2c^2} + \cdots$. But, if this mass is problematic, then so is the rest-mass energy $E = m_0c^2$ as we have argued throughout this work.

Mathematically, it is unclear how $E = m_0c^2$ from $E_{kinetic\ net} = m_{relativistic}\, c^2 - m_0c^2$ can be used to find the amount of the nuclear fission energy that can be released from the atom?

References

[1] Fitz G G 1889 Science 13 390
[2] Lorentz H 1892 Archives néerlandaises des sciences exactes et naturelles 25 363
[3] Einstein A 1905 Zur Electrodynamik bewegter Körpe, Annalen der Physik, On the Electrodynamics of Moving Bodies17 891-921

Journal of Physics: Conference Series **2793** (2024) 012003 doi:10.1088/1742-6596/2793/1/012003

Investigation of the weight effect on the oscillations of a magnetically-levitated body coupled to an energy harvester

A C Francis[1*], A Kandil[1,2], A ElSaid[1,3] and W K Zahra[1,4]

[1]Department of Mathematics, Institute of Basic and Applied Sciences, Egypt-Japan University of Science and Technology (E-JUST), New Borg El-Arab City, 21934, Alexandria, Egypt
[2]Department of Physics and Engineering Mathematics, Faculty of Electronic Engineering, Menouf, 32952, Menoufia University, Egypt
[3]Department of Mathematics and Engineering Physics, Faculty of Engineering, Mansoura University, Mansoura, Egypt
[4]Department of Engineering Physics and Mathematics, Faculty of Engineering, Tanta University, Tanta, Egypt

* Corresponding author's email: abraham.francis@ejust.edu.eg

Abstract. Energy harvesting technology utilizes magnetic oscillations to produce electrical energy which in turn powers any small device. This paper discusses the numerical analysis of a non-ideal magnetic levitation system, with emphasis on the energy harvesting potential considering the weight of the oscillating middle block. The system being considered is a magnetic levitation system connected to an electrodynamic shaker that excites the base of the system to produce oscillations. The outcome of modelling the Maglev system is a coupled highly nonlinear equation that resembles the duffing oscillator. Runge-Kutta method is carried out to describe the temporal behavior of the oscillatory middle block and hence the harvested power. The results of findings presented through phase planes, Poincare map and parametric variations indicate that varying parameters such as Ω and E, directly impact power harvesting and system stability. This paper explores periodic and non-periodic motions in the temporal behavior of the Maglev system, contributing to the current trend in energy harvesting. The novel addition of the middle block's weight offers crucial insights for enhancing energy harvesting devices, paving the way for future advancements in system design and application.

1. Introduction

Increase in research over the years that bothers on energy harvesting has been observed by curious minds because of the demand for alternate power source and global demand to reduce burning of fossil fuel for sustainable development. These energy harvesters include devices that generate energy from low power microelectromechanical systems [1]. These systems produce low powered energy used in turn to power electrical devices that are commonly used in remote areas where electric power is in scarce resource. With increase in demand for this technological knowhow, comes the interest in the analysis of nonlinear vibrating sources, piezoelectric materials which can be used to generate ambient power from surrounding vibrating source [2] and so on. Low level vibration sources for this smart device can be gotten from the environment, artificial vibrators, or oscillating devices [3]. This vibrating source can be divided into categories like the ambient vibration (structural and machinery) source,[1] that primarily originates from buildings, bridges, human activities, industrial machineries, and equipment.

Recent trends in research for ambient vibration is the oscillating device from a magnetically levitated object. The device involves an oscillating magnetic mass within two coils. This motion induces within the coil a difference in the magnetic field, subsequently generating an electric current through electromagnetic induction. In essence, the device harnesses the kinetic energy of the oscillating magnetic mass to produce electrical power through the fluctuation of magnetic fields within the coil [4].

For this levitation device to be self-sufficient and technologically friendly, it requires an excitation force to constantly base-excite it. This process of base exciting the Maglev system is referred to as the non-ideal excitation. Very few research has been conducted on the energy efficiency of a non-ideal Maglev system, we recognize that there are gaps in the study of non-ideal Maglev system. One such gap is considering the weight of the levitating magnet as this can influence the stability and enhance the system's energy harvesting potential.

This research focuses on using an electrodynamic shaker to base excite the energy harvester while considering the weight of the center magnet. Non ideal excitation sources are extensively discussed by [5–9], particularly recent works by [4, 7] are good research that relates to the proposed study. This paper presents an energy harvester that utilizes magnetic field force to make a central magnet oscillate while levitating. The aim is to inculcate the weight of the middle magnet in the model using nonlinear restoring forces.

This paper is structured thus: The next section details the figure of the Modelled system, the piezoelectric circuit for the energy harvesting, experimental data carefully derived from [4, 7], the governing equations, and how magnetic restoring forces are utilized to adjust the system's linear resonance. The third section explores the derivation of the modelled equation considering the mass of the middle block by axis translation principles. Following this, a set of detailed temporal profile plots is presented, to illustrate variations observed in the electrodynamic shaker's charge and the amplitude of the middle block. This aims to display the potential to harvest energy of the proposed system and examine the amplitude of the middle block. The last section outlines key features regarding the work and provides a concluding statement.

2. Mathematical model of the magnetic levitation mechanism

'Figure 1' illustrates the arrangement of the Maglev system connected to an energy harvesting apparatus. The design includes three magnetic blocks situated at the upper, central, and lower sections of the device. The central magnet experiences a downward displacement of magnitude Y_0, attributed to the influence of gravity. In this setup, the central magnetic block experience oscillations when the shaker induces vibrations, leading to changes in the magnetic field and consequently producing electrical current within the Maglev coils.

Moreover, the electrical shaker is stimulated by a voltage source with a harmonic waveform characterized by an amplitude e and frequency symbolized by Ω. The equations describing the system dynamics are defined as [7]

$$m\ddot{Y} + mg + F_{mag} + S\dot{q} + C_{me}\dot{Y} = 0, \tag{1}$$

$$L_s\ddot{Q} - S\dot{Y} + \frac{Q}{C_s} + R_s\dot{Q} = e\cos(\widehat{\Omega}t), \tag{2}$$

In the given context, where Y denotes the displacement of the middle block with mass m, Q is the electric charge in the shaker's circuit, C_{me} is the combined mechanical-electrical damping parameters, F_{mag} is the resultant magnetic force between the top and bottom magnets, S is the

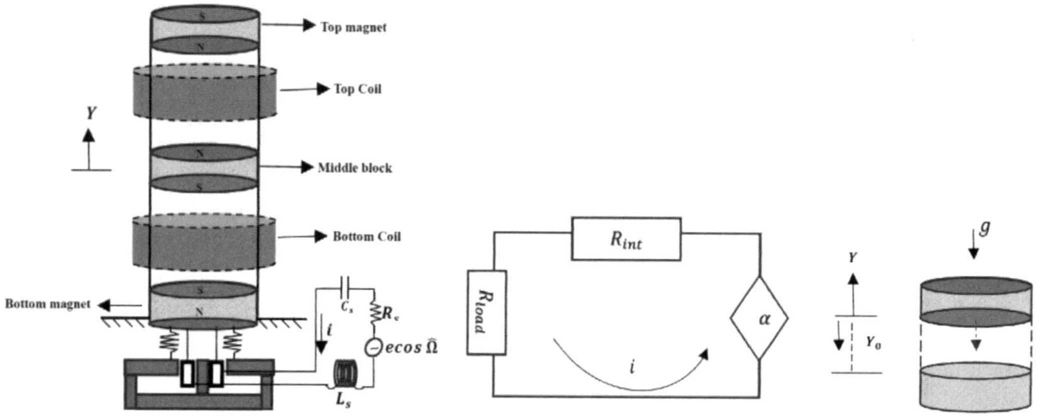

(a) (b) (c)

Figure 1. (a) Electro-dynamical Shaker-Driven Magnetic Levitation System (b) Energy harvesting circuit. (c) effect of considering weight on the middle block.

coupling parameter, and $F_{ex} = e\cos(\hat{\Omega}t)$ represents the external excitation or input voltage to the electrodynamic shaker. The magnetic force in the system is expressed as a power series up to the third order.

$$F_{mag} = F_b - F_t = \sum_{n=0}^{3} \alpha_n (d+Y)^n - \sum_{n=0}^{3} \alpha_n (d-Y)^n, \tag{3}$$

In this equation, F_b and F_t represent the forces exerted by the lower and upper magnets, respectively. The parameter d corresponds to the gap between each pair of magnets, and α_n are coefficients determined through experimental investigations of the magnetic forces. Therefore,

$$F_{mag} = k_1 Y + k_3 Y^3, \tag{4}$$

where $k_1 = 2\alpha_1 + 4d\alpha_2 + 6d^2\alpha_3$ and $k_3 = 2\alpha_3$. Therefore, equations (1) and (2) can be written as

$$m\ddot{Y} + mg + k_1 Y + k_3 Y^3 + S\dot{q} + C_{me}\dot{Y} = 0, \tag{5}$$

$$L_s\ddot{Q} - S\dot{Y} + \frac{Q}{C_s} + R_s\dot{Q} = e\cos(\hat{\Omega}t), \tag{6}$$

In the static equilibrium state, where the velocities $(\dot{Y}, \dot{Q})$ and accelerations $(\ddot{Y}, \ddot{Q})$ are zero, and the displacement $(Y) = -Y_0$. The weight of the middle block is substituted with,

$$mg = k_1 Y_0 + k_3 {Y_0}^3, \tag{7}$$

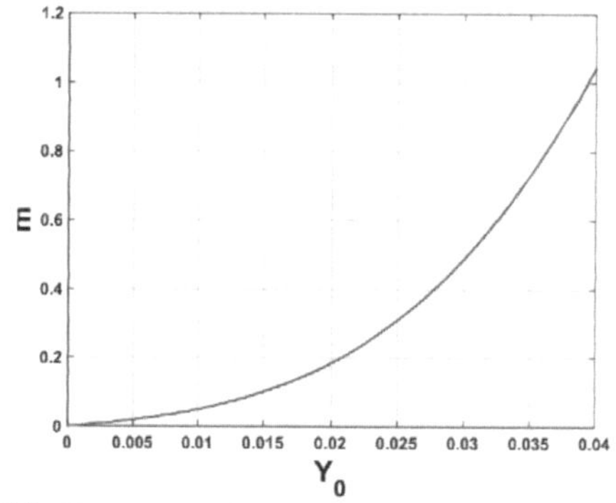

Figure 2: Nonlinear Relationship Between Mass and Initial Displacement in the Magnetic Levitation System.

A detailed inspection of equation (7) reveals in 'fig 2' that the mass of the middle block has an influence on the amplitude of the Maglev system. Based on the geometry of axis translation [11], we hence obtain:

$$Q = q \Rightarrow \dot{Q} = \dot{q} \Rightarrow \ddot{Q} = \ddot{q}, \tag{8a}$$

$$Y = y - Y_0 \Rightarrow \dot{Y} = \dot{y} \Rightarrow \ddot{Y} = \ddot{y}, \tag{8b}$$

Substituting equations (7) - (8) in (5) and (6) yields

$$\ddot{y} + c_{me}\dot{y} + \omega_1^2 y - \alpha_1 y^2 + \alpha_2 y^3 + \gamma_1 \dot{q} = 0, \quad (9a)$$
$$\ddot{q} + \eta\dot{q} + \omega_2^2 q - \gamma_2\, \dot{y} = E^* \cos(\hat{\Omega}t), \quad (9b)$$

where $\omega_1^2 = \frac{k_1+3k_3Y_0^2}{m}$, $c_{me} = \frac{C_{me}}{m}$, $\alpha_1 = \frac{3k_3Y_0}{m}$, $\alpha_2 = \frac{k_3}{m}$, $\gamma_1 = \frac{S}{m}$, $\gamma_2 = \frac{S}{L_s}$, $\eta = \frac{R_s}{L_s}$, $\omega_2^2 = \frac{1}{L_sC_s}$, $E^* = \frac{e}{L_s}$. Performing the normalization process using: $\tau = \omega_1 t, y = y^*(d - Y_0)$, $q = q^* q_0$, and removing the asterisk for convenience equation (9) yields

$$y'' + cy' + y - \eta_1 y^2 + \eta_2 y^3 + \eta_3 q' = 0, \quad (10a)$$
$$q'' + \delta q' + \omega^2 q - \eta_4 y' = E\cos(\Omega\tau), \quad (10b)$$

where $c = \frac{c_{me}}{\omega_1}$, $\eta_1 = \frac{\alpha_1(d-Y_0)}{\omega_1^2}$, $\eta_2 = \frac{\alpha_2(d-Y_0)^2}{\omega_1^2}$, $\eta_3 = \frac{\gamma_1 q_0}{\omega_1(d-Y_0)}$, $\delta = \frac{\eta}{\omega_1}$, $\omega^2 = \frac{\omega_2^2}{\omega_1^2}$, $\eta_4 = \frac{\gamma_2(d-Y_0)}{\omega_1 q_0}$, $E = \frac{E^*}{\omega_1^2 q_0}$, $\Omega = \frac{\hat{\Omega}}{\omega_1}$. The energy harvesting circuit equation is derived through the application of Kirchhoff's law for electrical circuits, as depicted in [7].

$$P = i^2 R_{load} = \left(\frac{c_e}{\alpha}\dot{Y}\right)^2 R_{load}, \quad (11)$$

where α denotes electromechanical interaction, i is the electric current and R_{load} is external resistance of the harvesting circuit. Upon subjecting it to the normalization analysis, we get

$$P_{dim} = (C_e y')^2, \quad (12)$$

where $C_e = \frac{c_e}{\alpha}(d - Y_0)\omega_1\sqrt{R_{load}}$.

3. Numerical results

The results got in experiments conducted by [4], concerning non-ideal model parameters identified in the theoretical studies outlined in [7], are provided as. $c = 0.2023$, $\eta_1 = 1.4289$, $\eta_2 = 3.0001$, $\eta_3 = 7.7866$, $\delta = 0.1318$, ω =2.6533, $\eta_4 = 0.04187$, $E = 0.0799$, $\Omega = 2.7444$,and $C_e = 0.0117$. Within this section, we conduct a comprehensive exploration of the Maglev system dynamics by employing numerical proceedings. Parametric variations, including Ω and E, are carried out to see the affect it would have on the system behaviour particularly on power harvesting and system stability.

'Figure 3' illustrates a periodic time-dependent profile of both the middle magnetic block and the charge of the electromechanical shaker at $E = 0.0799$ and $\Omega = 2.7444$. Even though the system presents a periodic motion, there is need to vary the system's amplitude (E) and frequency (Ω) of its voltage source to determine the right varied parameter for efficient energy harvest. At steady state motion the amplitude of the middle block is seen to be approximately 0.85 and harvested power of about 5.8×10^{-7}.

'Figure 4' illustrates the temporal profiles for both the power and velocity. Phase plane and Poincare map is further plotted in 'figure 5' to verify the periodic nature of the system. 'Figure 6-8' shows the variations in the parameters and their effect on the system under consideration. 'Figure 6'a-b indicates that as we vary the frequency of its voltage source from $\Omega = 2.6$ to 3.0, there is an increase in the average power output of the system. Indicating that the frequency of the shaker's voltage source has a direct relationship with the Maglev's power output regardless of the system stability. 'Figure 6(c)' to 5(d) shows how increase in the parameter that describes the amplitude of the shaker (E) has a slight but significant effect on the power harvested over time for constant value of $\Omega = 2.7444$.

Journal of Physics: Conference Series **2793** (2024) 012003 doi:10.1088/1742-6596/2793/1/012003

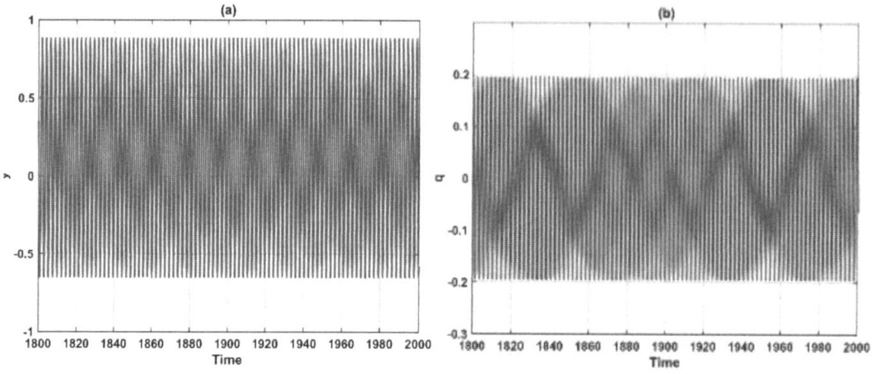

Figure 3. Temporal profiles of (a) the middle block movement and (b) electric shaker's charge at $E = 0.0799$

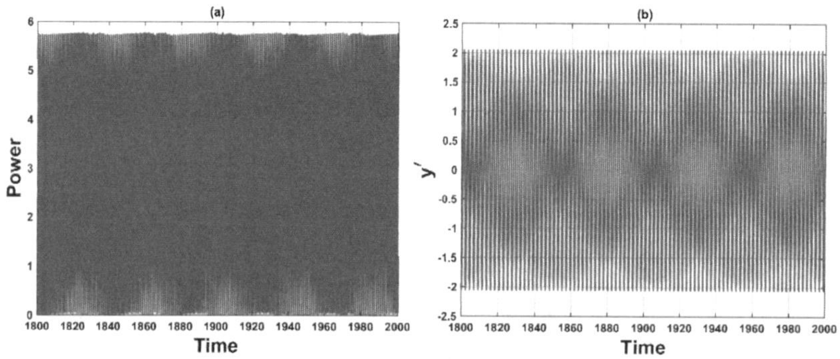

Figure 4. Temporal profiles of (a) the power generated at the top and bottom coil (b) the middle block's velocity generated at $E = 0.0799$

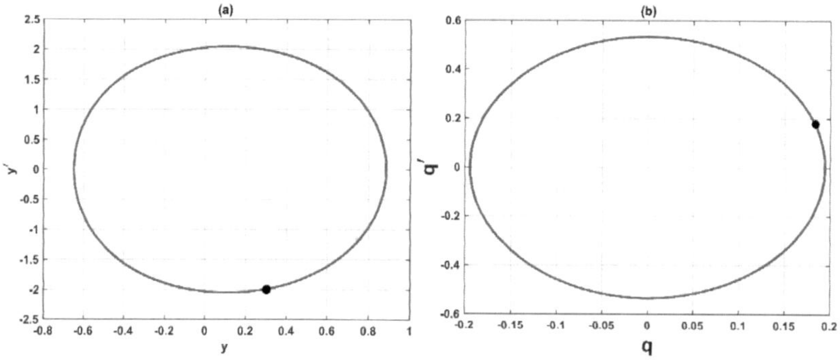

Figure 5. Phase plane (blue) and Poincare map (black) of middle block (a) and electric shaker's charge (b) for $E = 0.0799$

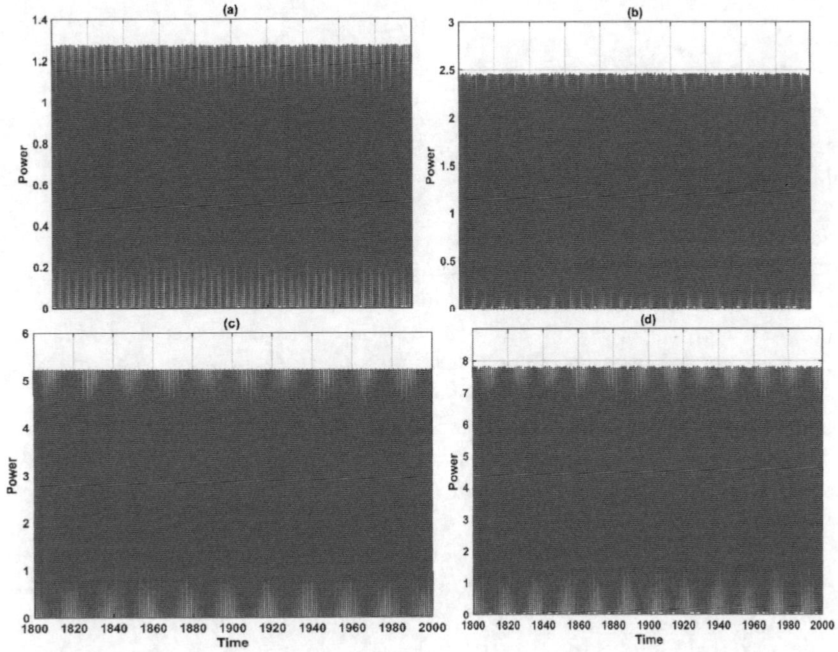

Figure 6. Temporal profiles of the power generated at the top and bottom coil (a) $\Omega = 2.6$ (b) $\Omega = 3.0$, for $E = 0.0779$ (c) $E = 0.0776$, (d) $E = 0.0868$ for $\Omega = 2.7444$

4. Conclusion

This present study focuses on a numerical analysis of a non-ideal magnetic levitation system connected to an electrodynamic shaker for base excitation, laying emphasis on considering the weight of the oscillating middle block. The energy harvesting device employs electromagnetic induction by harnessing the kinetic energy of an oscillating magnetic mass, producing electrical current through fluctuating magnetic fields within the coil. The system is modelled using a highly nonlinear equation resembling the Duffing oscillator, with numerical analysis validation using Runge-Kutta method. This paper explores the energy harvesting potential of the system through numerical investigations, phase planes, Poincare maps, and parametric variations. The novelty in this study is considering the weight of the oscillating middle block. The dynamics of the magnetic levitation system are explored by varying parameters such as frequency and amplitude of the voltage source, and their effects on oscillations, periodicity and power output are discussed. The paper discusses the relationship between the frequency of the shaker's voltage source and the power output, as well as the impact of amplitude variations of the middle block's vibration on the energy harvesting behaviour.

Acknowledgements

Abraham. C. Francis extends appreciation to JICA and EJUST, facilitated by TICAD7, for the research scholarship granted.

References

[1] Stephen, N. G. (2006). On energy harvesting from ambient vibration. *J. of Sound and Vibration*, 293 (1-2), 409–425. https://doi.org/10.1016/j.jsv.2005.10.003.

[2] Stanton, S. C., McGehee, C. C., & Mann, B. P. (2010). *Nonlinear dynamics for broadband energy harvesting*: Investigation of a bistable piezoelectric inertial generator. Physica D, 239 (10), 640–653. https://doi.org/10.1016/j.physd.2010.01.019.

[3] Kecik, K., Mitura, A., Lenci, S., & Warminski, J. (2017). Energy harvesting from a magnetic levitation system. *International Journal of Non-Linear Mechanics*, 94, 200–206. https://doi.org/10.1016/j.ijnonlinmec.2017.03.021.

[4] Mann, B. P., & Sims, N. D. (2009). Energy harvesting from the nonlinear oscillations of magnetic levitation. *J. of Sound and Vibration*, 319 (1-2), 515–530. https://doi.org/10.1016/j.jsv.2008.06.011

[5] Balthazar, J. M., Mook, D. T., Weber, H. I., Brasil, R. M., Fenili, A., Belato, D., & Felix, J. L. P. (2003). An overview on non-ideal vibrations. Meccanica, 38, 613-621.

[6] Iliuk, I., Balthazar, J. M., Tusset, A. M., Felix, J. L. P., & de Pontes, B. R. (2013). On Non-ideal and Chaotic Energy Harvester Behavior. *Differential Equations and Dynamical Systems*, 21 (1-2), 93–104. https://doi.org/10.1007/s12591-012-0127-5.

[7] Rocha, R. T., Balthazar, J. M., Tusset, A. M., de Souza, S. L. T., Janzen, F. C., & Arbex, H. C. (2019). On a non-ideal magnetic levitation system: nonlinear dynamical behavior and energy harvesting analyses. *Nonlinear Dynamics*, 95 (4), 3423–3438. https://doi.org/10.1007/s11071-019-04765-5

[8] Iliuk, I., Balthazar, J. M., Tusset, A. M., Piqueira, J. R. C., Rodrigues de Pontes, B., Felix, J. L. P., & Bueno, Á. (2013). A non-ideal portal frame energy harvester controlled using a pendulum. *European Physical Journal: Special Topics*, 222 (7), 1575–1586. https://doi.org/10.1140/epjst/e2013-01946-4.

[9] Bolla, M. R., Balthazar, J. M., Felix, J. L. P., & Mook, D. T. (2007). On an approximate analytical solution to a nonlinear vibrating problem, excited by a nonideal motor. *In Nonlinear Dynamics* 50(4), 841-847.

[10] Iliuk, I., Balthazar, J. M., Tusset, A. M., Felix, J. L. P., & de Pontes, B. R. (2012). Nonlinear dynamics and control strategies: on an energy harvester vibrating system with a linear form to non-ideal motor torque. *In MATEC Web of Conferences* (Vol. 1, p. 08003). EDP Sciences.

[11] Kandil, A. (2020). Investigation of the whirling motion and rub/impact occurrence in a 16-pole rotor active magnetic bearings system with constant stiffness. *Nonlinear Dynamics*, 102 (4), 2247-2265.

Journal of Physics: Conference Series **2793** (2024) 012004 doi:10.1088/1742-6596/2793/1/012004

Attempt to compensate for effects of energy conservation error on static pressure fluctuation using sub-grid scale components with the Reynolds number dependence of isotropic/anisotropic steady turbulence

Makoto Chitose[1], Hiroki Suzuki[2,*] and Toshinori Kouchi[2]

[1] Graduate School of Natural Science and Technology, Okayama University, 3-1-1 Tsushima-naka, Kita-ku, Okayama-shi, Okayama 700-8530, Japan
[2] Graduate School of Environmental, Life, Natural Science and Technology, Okayama University, 3-1-1 Tsushima-naka, Kita-ku, Okayama-shi, Okayama 700-8530, Japan

* Correspondig author's email: h.suzuki@okayama-u.ac.jp

Abstract. This study focuses on the kinetic energy conservation error and the compensation effect by adjusting a model constant. The kinetic energy conservation error is caused by the implicit time integration of the second-order Crank-Nicholson method, and we investigate how much this error affects the RMS of the static pressure fluctuation. Anisotropic and isotropic external force terms are used to generate a steady turbulent flow field, and the numerical analysis employs Large Eddy Simulation (LES) and the Smagorinsky model. Studies have shown that the influence of kinetic energy conservation errors on turbulent kinetic energy is limited, but large-scale turbulent structures are affected by errors in anisotropic turbulent fields with large Re numbers. Ta. Regarding the compensation of the RMS of the static pressure fluctuation, it was shown that the error could be reduced in both isotropic and anisotropic fields by adjusting the model constant, but there was a clear difference for the turbulent kinetic energy. This result shows that the compensation of kinetic energy conservation errors by a model constant is not perfect and that the errors can have different effects on the turbulence statistics.

1. Introduction

The flow around engineering equipment is often incompressible and turbulent. An accurate understanding of turbulent phenomena is widely required in engineering. Therefore, in addition to experimental measurements, numerical analysis is widely used to clarify turbulent phenomena. In numerical analysis, it is possible to determine quantities that are difficult to measure in experimental measurements [1]. Numerical analysis can also be used to determine quantities that are difficult to measure experimentally. Pressure is a necessary physical quantity in engineering, however it is known to be difficult to measure experimentally and high precision analysis is required to accurately determine the value of pressure in numerical analysis [2]. Numerical analysis uses direct numerical simulation, which solves all the equations directly, large eddy simulation (LES), which models small-scale eddies, and RANS, which models the entire flow field. DNS can provide highly accurate analysis, but requires a sufficiently fine computational grid, making it difficult to implement in engineering applications.

Numerical viscosity can affect the conservation laws in incompressible turbulence analysis. However, its absence does not guarantee compliance with these laws, especially if non-linear terms in the governing equations are incorrectly discretised [3]. Explicit time integration methods [4] are

Journal of Physics: Conference Series **2793** (2024) 012004 doi:10.1088/1742-6596/2793/1/012004

commonly used for spatial discretisation, while implicit methods [5] used in solvers such as OpenFOAM [6] can introduce energy conservation errors. These errors, evident in inviscid fields, affect static pressure fluctuations in turbulence [7.8] which are typically determined numerically due to measurement challenges. This study investigates the use of subgrid scale (SGS) dissipation to mitigate these errors and their impact on other turbulence statistics, in particular turbulent kinetic energy, which is known to be insensitive to energy conservation errors.

The aim of this study is to test whether the model constants can compensate for errors in the conservation of kinetic energy in analyses with errors in the conservation of kinetic energy. In analyses with errors in the conservation of kinetic energy, this study examines whether results comparable to those obtained in analyses without conservation errors can be obtained by changing the model constants. In the present analysis, the compensation of a model constant is investigated in terms of RMS static pressure fluctuations: in the analysis using LES, varying the model constant gives values of static pressure fluctuations of the same magnitude as those obtained using the DNS based on the spectral method and the results. The effect on the flow field of compensating the values of the model constants in terms of static pressure fluctuations is also investigated. It is investigated whether the turbulent kinetic energy is compensated in the same way when the model constants are used to compensate for static pressure fluctuations.

2. Numerical methods

2. 1. Details of the numerical simulation

The present analysis is based on three-dimensional incompressible turbulence [9]. The computational domain is a periodic cubic domain with one side of 2π to account for local turbulence. The governing equations are continuous with the Navier-Stokes (N-S) equations for a three-dimensional incompressible fluid, and an external force term is added to the N-S equation to generate a steady turbulent field. These equations have not been dimensionalised. The external force term is given by the linear forcing method [10-12]. The velocity component of the external force term is set to be either anisotropic or isotropic. The anisotropic external force term, f_x, f_y, f_z for x, y, z directions, respectively, is given by the following equation using analytical solutions based on previous studies [13].

$$f_x = -\cos(x)\sin(y), f_y = \sin(x)\cos(y) \text{ and } f_z = 0. \tag{1}$$

By combining these anisotropic external force terms in a way that maintains the above properties, an isotropic external force term is derived and is given by a previous study [7].

The computational grid is an equally spaced into 32 parts in the x, y and z directions. The spatial differencing method is the fourth order accurate central differencing method. The Smagorinsky model [1], one of the most commonly used models, is used as the SGS stress model of the LES. The Smagorinsky model requires the setting of the values of the model constants, which are generally defined by the model. It is known that the model constant C_s = 0.1. Changing this value will change the magnitude of the eddy viscous dissipation.

In previous studies, the use of the implicit time integration method is known to reduce the quadratic energy conservation characteristic when the quadratic energy conservation characteristic is explicit [6-8]. In this study, two types of time integration methods are used: explicit time integration and implicit time integration. For the explicit time integration method, the six-stage fourth-order accurate Runge-Kutta method is used [4]. The implicit time integration method used is the second-order accurate Crank-Nicolson method, which is one of the widely used methods in generic fluid analysis solvers.

The time evolution is from time step $t = 0$ to 3000. The Re numbers used in the calculations are set to 5, 10, 13, 20, 30, 50, 100, 200, 300 and 500; the model constants of the SGS model are set so that the

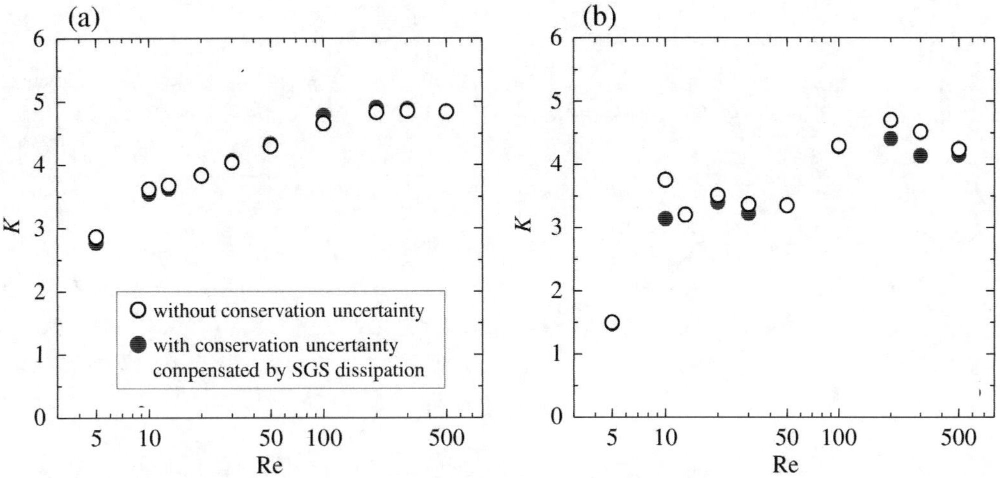

Figure 1. Effect of the conservation error on the dependence of the turbulent kinetic energy of steady turbulence on the Reynolds number (a) isotropic field (b) anisotropic field.

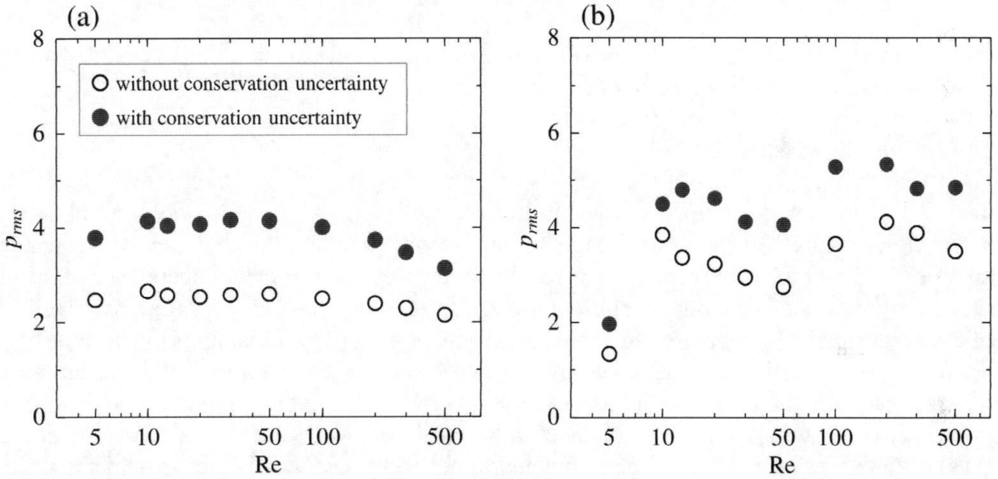

Figure 2. Effect of the conservation error on the dependence of the rms static pressure fluctuation on the Reynolds number (a) isotropic field (b) anisotropic field.

DNS results, performed at a sufficiently high spatial resolution, match the results of the RK4 method at Re = 300. For each of the isotropic and anisotropic external force fields, the calculated values of the model constants are: for the isotropic field, $C_s = 0.05733$, and for the anisotropic field, $C_s = 0.12$.

2.2. Verification on the conservation error setting

The current analysis first uses the inviscid flow field to investigate the kinetic energy conservation error introduced by the implicit time integration method. This field is created by omitting the viscous term from the governing equations and setting the turbulent viscous dissipation of the SGS model to zero. As a result, in the inviscid field, the kinetic energy analytically retains its initial value. To assess the conservation error, two different time integration methods are used to analyse the inviscid flow field. For conservation error free situations, the low memory, six stage, fourth order accurate Runge-Kutta method, an explicit time integration technique, is used. Conversely, the widely used second order accurate Crank-Nicolson method serves as the negative time integration method in fluid analysis solvers. The inviscid field evolves over time from $t = 0$ to 10. The initial velocity field, assumed to be isotropic,

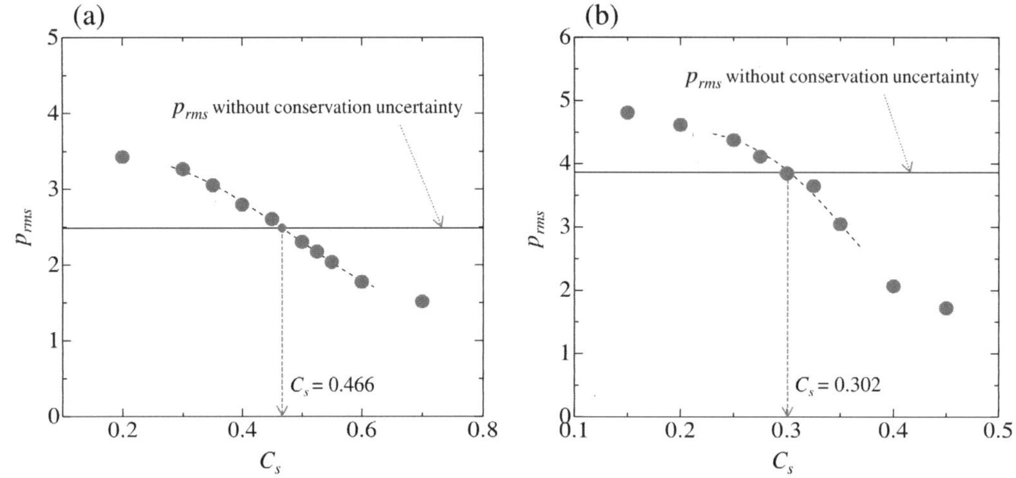

Figure 3. Compensation of static pressure fluctuation rms values affected by conservation errors by varying a model constant value. (a) isotropic field (b) anisotropic field.

has components similar to those of the isotropic external force and is normalised to ensure that the spatial mean of the kinetic energy is equal to one.

3. Results and discussion

From this point on, the results of the analysis using a steady turbulence field are presented. Figure 1 shows the dependence of the time-averaged turbulent kinetic energy on the Reynolds number in a steady turbulent field. Figure 1(a) shows the results for an isotropic field and Fig. 1(b) shows the results for an anisotropic field. For the isotropic field, the results for the Reynolds number dependence of the turbulent kinetic energy with and without kinetic energy are in good agreement. The time averaged turbulent kinetic energy tends to increase uniformly with increasing Reynolds number. Similarly, for the anisotropic field, the turbulent kinetic energy values obtained under conditions with conservation errors generally agree with the results without conservation errors. However, compared to the isotropic field, the discrepancy between the results obtained under conditions with and without conservation errors is larger. The time averaged values of the turbulent kinetic energy in the anisotropic field show a tendency to increase with increasing Reynolds number, with an extreme value. These results show that the turbulent kinetic energy is similar with and without conservation errors.

The study focuses on the RMS static pressure fluctuations in steady turbulence fields influenced by isotropic/anisotropic external force terms. Figure 2(a) shows the dependence of the RMS static pressure fluctuations on the Re number in a field generated by an isotropic external force, while Fig. 2(b) shows the results of an anisotropic external force. As shown in Fig. 2(a), the use of isotropic external forces results in higher RMS values of static pressure fluctuations under kinetic energy conservation error conditions than those without such errors. The RMS values show minimal sensitivity to variations in the Re number. Similarly, as observed in Fig. 2(b), the RMS value of the static pressure fluctuations increases when the anisotropic external force term is used due to kinetic energy conservation errors. The RMS value tends to increase and peaks at higher Re numbers. These results indicate that kinetic energy conservation errors increase the RMS value of static pressure fluctuations, regardless of whether the conditions are isotropic or anisotropic.

The value of the RMS static pressure variation is compensated by model constants. As previously discussed, the value of the static pressure fluctuation RMS increases due to the effect of kinetic energy conservation errors, therefore in analyses with conservation errors, the value is compensated by varying

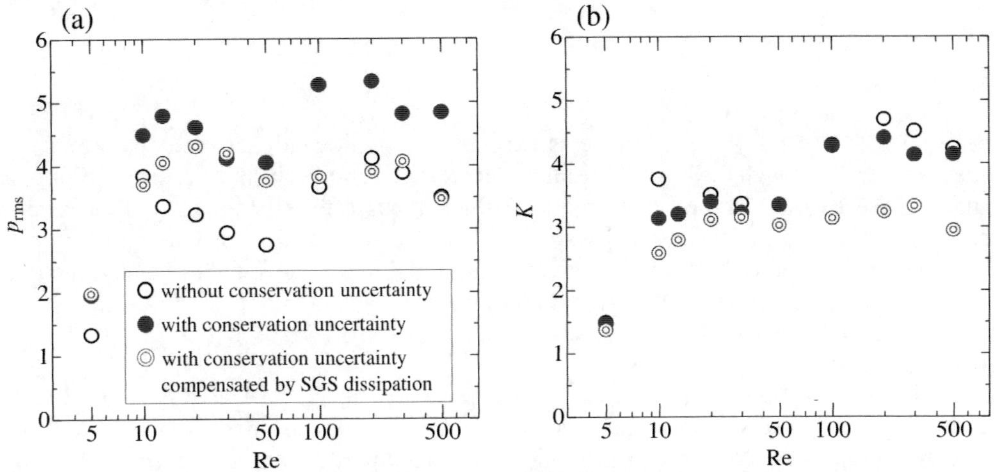

Figure 4. Reynolds number dependence of the anisotropic turbulence field with compensated conservation errors focusing on the rms static pressure fluctuation (a) rms static pressure fluctuation (b) turbulent kinetic energy.

the model constant. The Runge-Kutta and Crank-Nicolson methods are used to investigate the compensation using the value of the static pressure fluctuation at Re = 300. Under conditions where the results of the Runge-Kutta method are free from conservation errors, and by varying the model constant in the analysis using the Crank-Nicolson method, a model constant is found for which the value of the static pressure fluctuation RMS agrees with the value of the Runge-Kutta method. Figure 3 (a) shows the dependence of the RMS static pressure variation on the value of the model constant in an isotropic external force field, while Fig. b shows it in an anisotropic field. As shown in Fig.3 (a), changing the value of the model constant can cause the RMS value of the static pressure fluctuation to overlap with results without conservation errors. Similarly, in Fig. 3 (b), the Re-number dependence of the RMS value of the static pressure fluctuation has an intersection with the result without conservation error. The value of the model constant at the intersection is C_s = 0.466 for isotropic fields and C_s = 0.302 for anisotropic fields.

The study validates compensation using model constants based on RMS static pressure fluctuation values at Re = 300 through steady-state turbulence analysis. Three scenarios are compared: conditions with kinetic energy conservation errors, conditions with conservation errors but compensated by model constants, and conditions without conservation errors. Figure 4(a) shows the dependence of the time-averaged RMS static pressure fluctuation on the Re number, while Fig. 4(b) shows the dependence of the time-averaged turbulent kinetic energy on the Re number. As can be seen in Fig. 4(a), in the region of Re = 100 or higher, results under conservation error conditions compensated with model constants perform better than those without conservation errors. However, at lower Reynolds numbers, even with modified model constants, the RMS values of the static pressure fluctuations under conservation error conditions differ significantly from those without conservation errors. For turbulent energy, as shown in Fig. 4(b), compensation with model constants leads to a significant deviation from the results without conservation errors in high Reynolds number regions. In contrast, at lower Reynolds numbers the discrepancy between the modified model constant results and those without conservation errors is small.

4. Conclusion

This study investigates the possibility of compensating for static pressure fluctuation errors in kinetic energy conservation error analyses using model constants. Kinetic energy errors introduced by the Crank-Nicolson method are contrasted with the error-free six-stage fourth-order Runge-Kutta method. Turbulence is simulated in a periodic cube using linear forcing to add anisotropic and isotropic external forces. The governing equations are discretised using a fourth-order central differencing method, preserving quadratic quantities. Large vortex simulation and the Smagorinsky model are used. The analysis shows that the turbulent kinetic energy is not significantly affected by kinetic energy conservation errors, although the large-scale turbulence structure varies in anisotropic conditions at high Re numbers. The RMS values of static pressure fluctuations affected by conservation errors were analysed by adjusting the model constants to align the RMS values in both error and error-free conditions. Although the static pressure fluctuations were aligned in high Re regions, the turbulent kinetic energy showed significant deviations from the error-free results.

Acknowledgments
The author is deeply grateful to Prof. Shinsuke Mochizuki (Yamaguchi University) and Mr. Ryoma Honda (Yamaguchi University) for their support in carrying out this work. This work was partly supported by the Japanese Ministry of Education, Culture, Sports, Science and Technology through Grants-in-Aid (Nos. 21K03859 and 22H01684).

References
[1] Pope S B 2000 Turbulent Flows. Cambridge University Press
[2] Hirabayashi R, Suzuki H, Mochizuki S 2023 *J. Phys.: Conf. Ser.* 2585 012006 (14 pages).
[3] Morinishi Y, Lund T S, Vasilyev O V, Moin P 1998 *J. Comput. Phys.* 143 90-124
[4] Suzuki H, Hasebe K, Hasegawa Y, Ushijima T 2020 *J. Phys.: Conf. Ser.* 1633 012021 (9 pages)
[5] Ham F E, Lien F S, Strong A B 2002 *J. Comput. Phys.* 177 117-133
[6] Komen E, Shams A, Camilo L, Koren B 2014 *Comput. Fluids* 96 87-104
[7] Honda R, Suzuki H, Mochizuki S 2021 *J. Phys.: Conf. Ser.* 2090 012145 (8 pages)
[8] Honda R, Suzuki H, Mochizuki S 2021 *J. Phys.: Conf. Ser.* 1978 012026 (7 pages)
[9] Suzuki H, Nagata K, Sakai Y, Hayase T, Hasegawa Y, Ushijima T 2013 *Int. J. Numer. Meth. Fluids* 73 509-522
[10] Lundgren T S 2003 *Ann Res. Briefs of Center for Turb. Res.* 461-473
[11] Rosales C and Meneveau C 2005 *Phys. Fluids* 17 095106
[12] Carroll P L, Blanquart G 2013 *Phys. Fluids* 25 105114
[13] Goto S, Vassilicos J C 2015 *Phys. Lett. A* 379 1144-1148

Journal of Physics: Conference Series **2793** (2024) 012005 doi:10.1088/1742-6596/2793/1/012005

Analytical Fermi energy for broadened Landau levels of 2DEG under a perpendicular strong magnetic field

L J Sese*, **C J Bercasio**, **R Gammag**

Department of Physics, Mapúa University, Intramuros, Manila, Philippines

E-mail: `ljfsese@mapua.edu.ph`*

Abstract. An analytical form of the Fermi energy is derived for broadened Landau levels (LLs) of the two-dimensional electron gas (2DEG) under a perpendicular strong magnetic field. A Gaussian density of states with the broadening parameter Γ is used to derive the analytical form under the assumption of asymptotically non-overlapping states between two consecutive Landau levels, i.e. $\Gamma/\hbar\omega_c \to 0$. It is shown that the smoothening of the Fermi energy around the neighborhood for fully filled Landau levels is due to the linear dependence of Γ and its factor with the inverse error function. The validity of the analytical form shows a threshold up to $\Gamma/\hbar\omega_c \sim 0.23$.

1. Introduction

A current-carrying two-dimensional conductor placed in a perpendicular magnetic field B will have a transverse voltage - a Hall voltage [1]. The Hall effect emerges from the Lorentz force, where charged particles accumulate on one side of the conductor, thus producing Hall voltage transverse to the direction of the electric current.

When the applied magnetic field is strong, and the sample is further subject to low temperature T, a two-dimensional electron gas (2DEG) manifests pronounced quantization of its Hall resistance. This observation is named the quantum Hall effect (QHE) [2].

The emergence of QHE is due to the quantization of energy levels - the so-called Landau levels (LLs) [3]. The oscillation of thermodynamic quantities are also attributed to LLs [4–7]. The Landau levels of an ideal 2DEG are represented in the density of states (DOS) as a series of Dirac-delta distributions with gaps proportional to B,

$$D(E) = \frac{2eB}{h} \sum_n \delta(E - E_n). \tag{1}$$

The factor $2eB/h$ describes the degeneracy per LL, where e is the electron charge and h is the Planck's constant. Here, $E_n = (n + 1/2)\hbar\omega_c$ is the nth-LL, where $\hbar$ is the reduced Planck's constant. The cyclotron frequency is defined as $\omega_c = eB/m^*$ for a given effective mass m^*. Several thermodynamic quantities such as the magnetization [8] and the heat capacity [9] have already been derived for an ideal 2DEG under a perpendicular B. These thermodynamic quantities are functions of B and T.

Nevertheless, it is observed that LLs are broadened from the extracted DOS inferred from magnetocapacitance measurement [10, 11], magnetization [6, 7, 12], and heat capacity [4, 5].

Journal of Physics: Conference Series **2793** (2024) 012005 doi:10.1088/1742-6596/2793/1/012005

A broadened LL means that several energy states exist around the neighborhood of a LL. The broadening is attributed to disorders that are present such as defects, impurities, and other inhomogeneities in the system [10]. Shapes of DOS with broadened LL are fitted either with a Lorentzian, Gaussian, or a semielliptical distribution function [4, 6, 7]. An example of the fitting procedure is from the study of Gornik et al. [4], where it yields a DOS with Gaussian peaks. The Gaussian-shaped DOS takes form [3, 10, 13–15],

$$D(E) = \frac{2eB}{h}\sum_n \left(\frac{1}{2\pi\Gamma^2}\right)^{1/2} \cdot \exp\left[-\frac{(E-E_n)^2}{2\Gamma^2}\right]. \tag{2}$$

Here, Γ is known as the broadening parameter. In the limiting case of $\Gamma \to 0$, the Gaussian DOS will take its ideal Dirac-delta DOS in Eq. (1).

This current study aims to obtain, for the first time, an analytical form of the Fermi energy for a broadened LL. This is done by utilizing the one-LL approximation, which assumes a partially-filled LL with asymptotically no overlapping of states between two consecutive LLs. In addition, the validity of the analytical form will also be determined by the ratio of $\Gamma/\hbar\omega_c$. Spin-splitting is not considered in the derivation.

The use of Eq. (2) is necessary to calculate different thermodynamic quantities encapsulating the effect of Γ. However, we need to obtain the chemical potential prior to these calculations. The chemical potential is defined as the energy required to add another particle. This is determined by fixing the electron concentration N using the equation $N = \int_0^{+\infty} D(E)f(E)\mathrm{d}E$. Here, $f(E) = 1/(1+\exp[(E-\mu)/k_BT])$ is the Fermi-Dirac (FD) distribution and k_B the Boltzmann constant. The study of Vagner et al. [8], already provides the exact form of the chemical potential for varying B and T. However, complications arise when utilizing a Gaussian DOS. From the recent literature, we can only calculate $\mu(B,T)$ by means of the numerical root-finding method as it was already established in the study of Gammag and Villagonzalo [15].

2. Analytical Fermi energy

As the system's temperature approaches 0 K, the electrons occupy the minimum energy states obeying the Pauli exclusion principle. The last energy state occupied by the electron is what we call the Fermi energy μ_0. This is also the chemical potential at $T = 0$ K. We can calculate the Fermi energy as the upper limit in the integral by considering a constant electron concentration,

$$N = \int_0^{\mu_0} D(E)\,\mathrm{d}E. \tag{3}$$

To consider a system of 2DEG with broadened LLs, we substitute a Gaussian-shaped DOS in the integral.

When there is no overlap between two consecutive LLs, the integral will have the form

$$N = \frac{2eB}{h}(n'+1) - \frac{2eB}{h}\int_{\mu_0}^{+\infty}\left(\frac{1}{2\pi\Gamma^2}\right)^{1/2}\exp\left[-\frac{(E-E_{n'})^2}{2\Gamma^2}\right]\,\mathrm{d}E. \tag{4}$$

Here, n' corresponds to the last LL that is partially filled. We can determine n' as the integer value from the ratio of the total electron concentration per LL degeneracy. This is expressed as $\lfloor hN/2eB \rfloor$, where $\lfloor x \rfloor$ is the floor function of x. The first term of Eq. (4) provides the electron concentration for the fully-filled LLs up to $E_{n'}$. If $E_{n'}$ is partially filled, we subtract the remainder as an integral with the lower limit, μ_0. The integral in the last term can be evaluated utilizing the complementary error function, $\mathrm{erfc}(z) = \int_z^{+\infty}(2/\sqrt{\pi})e^{-t^2}\,dt$ [16]. Using the identity, $\mathrm{erfc}(z) = 1 - \mathrm{erf}(z)$, Eq. (4) yields

$$N = \frac{2eB}{h}\left(n'+\frac{1}{2}\right) + \frac{eB}{h}\mathrm{erf}\left(\frac{\mu_0 - E_{n'}}{\sqrt{2}\Gamma}\right), \tag{5}$$

Journal of Physics: Conference Series **2793** (2024) 012005 doi:10.1088/1742-6596/2793/1/012005

where $\mathrm{erf}(z) = \int_0^z (2/\sqrt{\pi})e^{-t^2}\,dt$ is the error function. We can take the inverse of the error function to obtain the form of μ_0 explicitly [17, 18]. This gives

$$\mu_0(B) = \left(\left\lfloor \frac{hN}{2eB} \right\rfloor + \frac{1}{2}\right)\frac{heB}{2\pi m^*} + \sqrt{2}\Gamma\,\mathrm{erf}^{-1}\left[\frac{hN}{eB} - 2\left(\left\lfloor \frac{hN}{2eB} \right\rfloor + \frac{1}{2}\right)\right]. \tag{6}$$

3. Results and discussion

The result in Eq. (6) can be expressed in terms of the filling factor ν. The filling factor quantifies the number of LLs that are already occupied, i.e., $\nu = hN/eB$. Rewriting μ_0 in terms of ν takes the form,

$$\mu_0(\nu) = \left(\left\lfloor \frac{\nu}{2} \right\rfloor + \frac{1}{2}\right)\frac{2E_F}{\nu} - \sqrt{2}\Gamma\,\mathrm{erf}^{-1}\left(2\left\lfloor \frac{\nu}{2} \right\rfloor + 1 - \nu\right), \tag{7}$$

where $E_F = h^2N/4\pi m^*$, is the Fermi energy of 2DEG without an applied B. Here, $\mathrm{erf}^{-1}(z)$ is the inverse error function and it has a domain of $z \in (-1, 1)$ [17, 18].

The first term of $\mu_0(\nu)$ shows the ideal Fermi energy of 2DEG under a perpendicular B. Hence, we can obtain the ideal case for $\Gamma \to 0$. This is equal to the first term of the chemical potential that is obtained by Vagner et al. in Ref. [8]. The first term of $\mu_0(\nu)$ is responsible for the sawtooth oscillation of the chemical potential. The floor function can relate to a sawtooth function by $\lfloor x \rfloor = \{x\} - x$, where $\{x\}$ is the sawtooth function.

Considering the case for half-filling $\nu = 2m + 1$, where m is an integer, the floor function $\lfloor (2m+1)/2 \rfloor$ reduces to m. Substituting this in Eq. (7) retrieves E_F since $\mathrm{erf}^{-1}(0) = 0$. The form in Eq. (7) recovers the case for half-filling ν.

The effect of broadening in $\mu_0(\nu)$ is seen in the second term. Mathematically, it is linear in Γ. Considering non-integer ν, the sharp sawtooth is smoothened due to the inverse error function. Specifically, for $2m < \nu < 2m+1$, $\mu_0(\nu)$ is less than the ideal case. This is due to the additional states that can exist below the energy of the last LL. Meanwhile, for $2m+1 < \nu < 2m+2$, $\mu_0(\nu)$ becomes greater than the ideal case. Due to the broadening, energy states can exist greater than the energy of the last LL.

The result in Eq. (7) is also verified by having a numerical simulation from the method implemented in Ref. [15]. This is shown in Fig. 1 as a superimposed plot between the two methods. Here, μ_0 is obtained from Eq. (3) by a root-finding algorithm. This is done by fixing and setting N equal to 3.6×10^{11} cm^{-2}. This value of N is of the same order of magnitude as in experimental studies for GaAs-AlGaAs heterostructure [4–6, 10, 12]. The value for m^* is set to $0.0667 m_e$, where m_e is the mass of the electron. We set a constant broadening, $\Gamma = 0.2$ meV. These parameters are also used to obtain the plot for Eq. (7). In addition, this is plotted with respect to the ratio of the broadening and the level spacing $\Gamma/\hbar\omega_c$ to check the validity of the approximation in obtaining Eq. (7).

In the regime of strong magnetic field, i.e., $\Gamma/\hbar\omega_c \ll 1$, the analytical form and the numerical simulation are the same. This shows the validity of the approximation upon the said regime. Since the level spacing is much greater than the broadening, then the overlapping states of a Gaussian DOS between two consecutive LLs become asymptotically small. The validity of the approximation shows a threshold up to $\Gamma/\hbar\omega_c \sim 0.23$ as shown in the inset of Fig. 1. This implies that the overlapping states between two consecutive LLs become prominent. There will be a double-counting of states upon using Eq. (4), which leads to a significant error comparable to the numerical simulation. It is known that $\mu_0 \to E_F$ as $B \to 0$.

4. Conclusion

Using the one-LL approximation enables us to obtain an analytical form for the Fermi energy μ_0 of 2DEG in a perpendicular strong B. The approximation is valid when there is no overlap

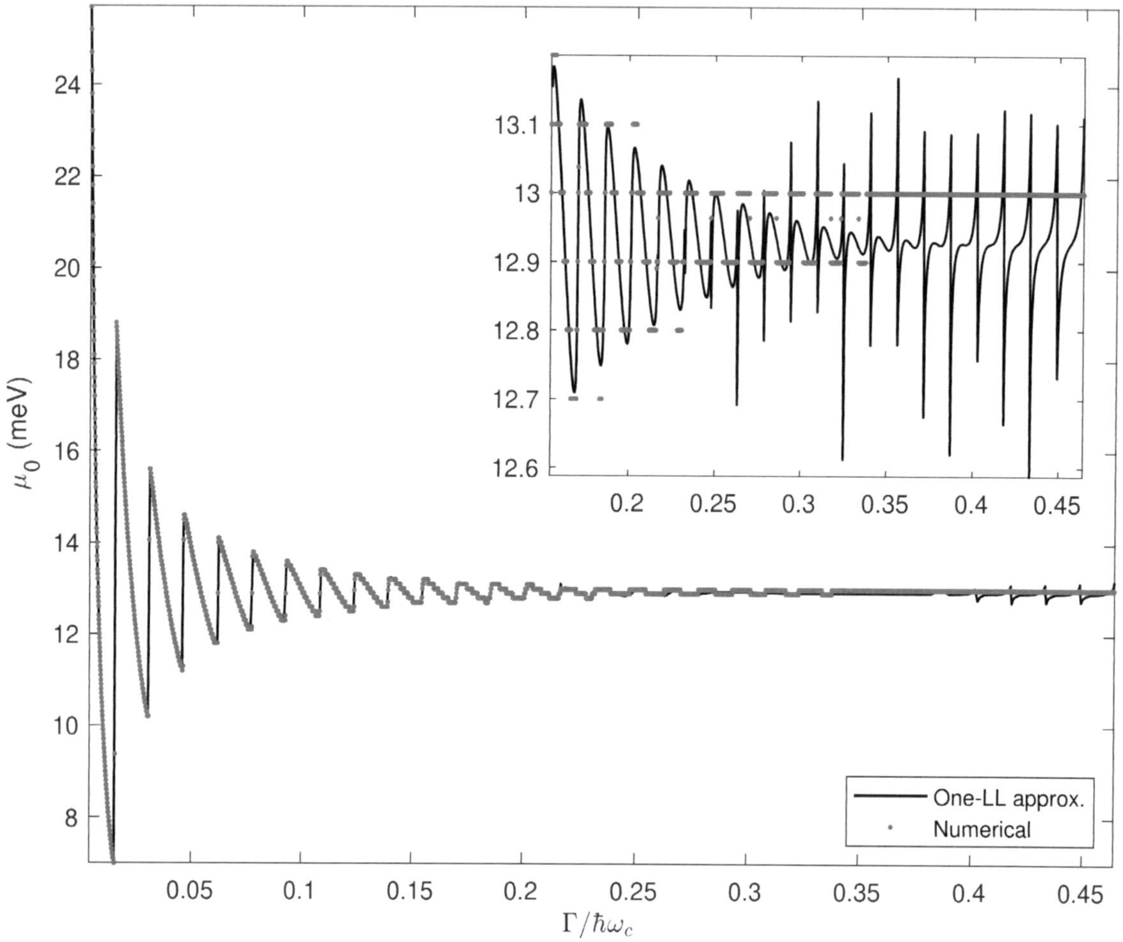

Figure 1: Superimposed plot of the Fermi energy of 2DEG with broadened LLs ($\Gamma = 0.2$ meV) for the analytical form (continuous line) and numerical simulation (dotted lines). The inset shows the magnified superimposed plot for $0.16 < \Gamma/\hbar\omega_c < 0.46$.

between LLs. The derived form in Eq. (7) can retrieve the case for an ideal 2DEG, i.e., $\Gamma \to 0$. Moreover, we can directly obtain from the derived form the Fermi energy without an applied B for the half-filling ν. The validity of the approximation is satisfied for $\Gamma/\hbar\omega_c \ll 1$, i.e., the regime for a strong magnetic field. The approximation works up to a threshold of $\Gamma/\hbar\omega_c \sim 0.23$.

Despite the limitations, Eq. (7) still remain relevant since most of the interesting effects (e.g. QHE, de Haas-van Alphen effect (dHvA), Shubnikov-de Haas effect (SdH)) happen in the regime of strong magnetic field [7, 12, 19], where the LL spacing is more prominent than Γ. The derivations herein can also be used for studying the effect of broadening that depends on B as long as the condition for strong magnetic field regime is satisfied.

References

[1] E. Hall, On a new action of the magnet on electric currents, *Am. J. Math.* **2**, 287 (1879).

[2] T. Ando, Y. Matsumoto, and Y. Uemura, Theory of Hall effect in a two-dimensional electron system, *J. Phys. Soc. Japan* **39**, 279 (1975).

Journal of Physics: Conference Series **2793** (2024) 012005 doi:10.1088/1742-6596/2793/1/012005

[3] T. Ando, A. Fowler, and F. Stern, Electronic properties of two-dimensional systems, *Rev. Mod. Phys.* **54**, 437 (1982).

[4] E. Gornik, R. Lassnig, G. Strasser, H. Störmer, A. Gossard, and W. Wiegmann, Specific heat of two-dimensional electrons in GaAs-GaAlAs multilayers, *Phys. Rev. Lett.* **54**, 1820 (1985).

[5] J. Wang, D. Tsui, M. Santos, and M. Shayegan, Heat-capacity study of two-dimensional electrons in GaAs/$Al_xGa_{1-x}As$ multiple-quantum-well structures in high magnetic fields: Spin-split landau levels, *Phys. Rev. B* **45**, 4384 (1992).

[6] M. Zhu, A. Usher, A. J. Matthews, A. Potts, M. Elliott, W. G. Herrenden-Harker, D. A. Herrenden-Harker, and M. Y. Simmons, Magnetization measurements of high-mobility two-dimensional electron gases, *Phys. Rev. B* **67**, 155329 (2003).

[7] M. Wilde, M. Schwarz, C. Heyn, D. Heitmann, and D. Grundler, Experimental evidence of the ideal de Haas–van Alphen effect in a two-dimensional system, *Phys. Rev. B* **73**, 125325 (2006).

[8] I. Vagner, T. Maniv, and E. Ehrenfreund, Ideally conducting phases in quasi two-dimensional conductors, *Phys. Rev. Lett.* **51**, 1700 (1983).

[9] C. Villagonzalo and R. Gammag, The intrinsic features of the specific heat at half-filled Landau levels of two-dimensional electron systems, *J. Low Temp. Phys.* **163**, 43 (2011).

[10] T. Smith, B. Goldberg, P. Stiles, and M. Heiblum, Direct measurement of the density of states of a two-dimensional electron gas, *Phys. Rev. B* **32**, 2696 (1985).

[11] V. Mosser, D. Weiss, K. Klitzing, K. Ploog, and G. Weimann, Density of states of GaAs-AlGaAs-heterostructures deduced from temperature dependent magnetocapacitance measurements, *Solid State Commun.* **58**, 5 (1986).

[12] J. Eisenstein, H. Stormer, V. Narayanamurti, A. Cho, A. Gossard, and C. Tu, Density of states and de Haas–van Alphen effect in two-dimensional electron systems, *Phys. Rev. Lett.* **55**, 875 (1985).

[13] W. Zawadzki and R. Lassnig, Specific heat and magneto-thermal oscillations of two-dimensional electron gas in a magnetic field, *Solid State Commun.* **50**, 537 (1984).

[14] W. Zawadzki and R. Lassnig, Magnetization, specific heat, magneto-thermal effect and thermoelectric power of two-dimensional electron gas in a quantizing magnetic field, *Surf. Sci* **142**, 225 (1984).

[15] R. Gammag and C. Villagonzalo, The interplay of Landau level broadening and temperature on two-dimensional electron systems, *Solid State Commun.* **146**, 487 (2008).

[16] I. Gradshteyn and I. Ryzhik, *Table of Integrals, Series, and Products (Eighth Edition)* (Academic Press, Boston, USA, 2015).

[17] J. Philip, The function Inverfcθ, *Aust. J. Phys.* **13**, 13 (1960).

[18] L. Carlitz, The inverse of the error function, *Pac. J. Math* **13**, 459 (1963).

[19] G. S. Kliros, Effect of the Landau level broadening on the quantum Hall conductance, *Il Nuovo Cimento D* **13**, 99 (1991).

Journal of Physics: Conference Series **2793** (2024) 012006 doi:10.1088/1742-6596/2793/1/012006

Heating opaque material sheets by a blue laser

Kwanruthai Sibsiri, Puchong Kijamnajsuk, Malee Sutthiopad and Chaiya Luengviriya*

Department of Physics, Kasetsart University, 50 Phaholyothin Road, Jatujak, Bangkok 10900, Thailand

* Corresponding author's E-mail: fscicyl@ku.ac.th

Abstract. We present a study of heating small opaque substrates using a blue laser diode. It can be used as a method of temperature control for small sensors, e.g., atomic magnetometers. A blue laser from a low-cost engraving machine is used to illuminate an opaque material sheet ($1\times1\times0.2$ cm^3) attached to a 1-cm glass cuvette whose size is similar to a typical vapor cell in compact atomic magnetometers. The input electric power of the laser diode (P) is supplied by a laser driver board and controlled via a computer by a pulse width modulation. We investigate the heating method using the electric power (P) between 2.67 and 4.67 W and 4 types of opaque materials: a silica bonded investment material, a gypsum bonded investment material, a silicone epoxy, and a mortar. The results show that in all cases, the temperature of the cuvette increases with time until it reaches a maximum value (T_{max}) within 10 mins. For all tested materials, T_{max} always increases with the power P. For a given power P, T_{max} depends on the material type. At the maximum power of 4.67 W, T_{max} are 52°C, 64°C, 81°C, and 88°C for the silica bonded investment material, the gypsum bonded investment material, the silicone epoxy, and the mortar, respectively. Therefore, among the tested materials, mortar is the most effective material for laser heating.

1. Introduction

Blue laser is used in many applications such as annealing, drilling, cutting, welding, alloying, and others [1-4]. When a substrate surface is illuminated by a laser beam, the beam energy is partially absorbed and the substrate undergoes solid heating, melting, or evaporation depending on many parameters including the laser power and the reflexivity of substrate [1]. In wearable piezoresistive sensors for detecting pressure changes, the microstructures of sensing layers are formed by blue laser annealing on carbon nanotube layers [2]. In electronic industries, welding thin foils of the so-called difficult-to-weld metals is successfully done by using blue laser [3, 4].

In this article, we demonstrate heating of small opaque substrates using a blue laser. It can be used to control the temperature of small sensors such as atomic magnetometers (AM). These sensors are very sensitive to magnetic fields and find many applications in biomagnetism such as magnetoencephalography, magnetocardiography, nuclear magnetic resonance and magnetic resonance imaging [5-8]. AM works by optical interrogation of orientation of alkali-atom spins in a vapor cell and its sensitivity depends on many factors including the density of alkali atoms which increases with the temperature [5,9]. The vapor cell can be heated by many methods such as hot air, hot water, AC electrical current and laser heating [10]. Among these methods, the laser heating does not cause magnetic field interference and consumes little energy. Examples of laser heating usages are as follows. A 940-nm laser was used for heating a 1-cm^3 vapor cell in a compact atomic magnetometer [11]. A micromachined alkali vapor cell with a volume of 1.50-18 mm^3 was heated by using a 1.5-μm laser [12-14]. The temperature of a 1-mm^3 microelectromechanical vapor cell is increased by using 915-nm laser [15].

Here, we present a study of heating opaque substrates by using a blue laser diode, a component of a low-cost engraving machine. A 1-cm glass cuvette is used to mimic a vapor cell of a compact AM. The laser illuminates an opaque material sheet attached to the cuvette. The input electric power of the laser (P) and the opaque material are varied.

2. Methods

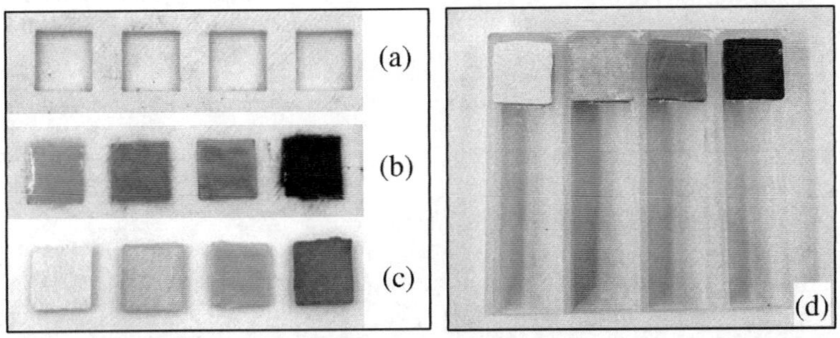

Figure 1. Preparation of opaque material sheets: (a) molds with dimensions of 1x1x0.2 cm^2, (b) mixtures of material powder filled in the molds, (c) hard and dry sheets after 24 hours, and (d) the opaque sheets attached on 1-cm glass cuvettes before laser heating. From left to right, materials are a silica bonded investment material, a gypsum bonded investment material, a silicone epoxy, and a mortar.

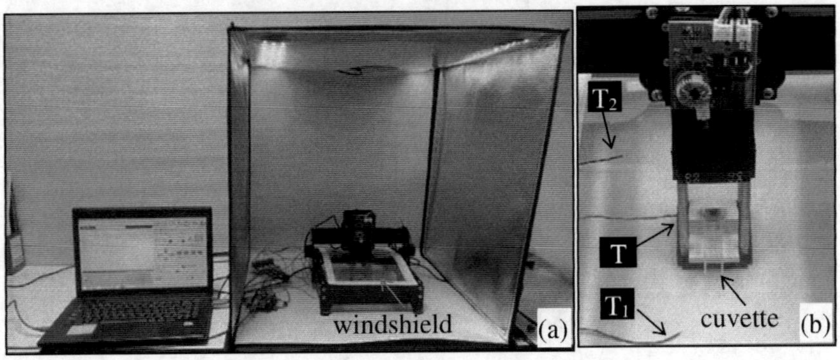

Figure 2. Laser heating setup. (a) An engraving blue laser in an optical enclosure is controlled by a computer. (b) The blue laser illuminates an opaque material sheet attached on a glass cuvette. Thermocouples measure the temperatures of the opaque material sheet (T) and environment under (T_1) and upper (T_2) transparent windshield.

As shown in figure 1, we prepare opaque material sheets with different materials: a silica bonded investment material, a gypsum bonded investment material, a silicone epoxy, and a mortar. The material powder is mixed well with a small portion of water and put into a mold with dimensions of 1×1×0.2 cm^3. After 24 hours the mixtures are hard and dry, so they are removed from the molds. The opaque material sheet is attached to a glass cuvette (a 1-cm spectrophotometric cell) using a heatsink plaster and ready for experiment.

The laser heating setup is shown in figure 2(a). A blue laser beam from a low-cost engraving machine is used for heating the opaque sheet. For safety purposes, the laser set is placed in an optical enclosure whose windows are closed during the experiments. The input electric current of the laser diode is supplied by a laser driver board which is controlled via a computer. A pulse width modulation (PWM) technique is used to control the electric current and the corresponding input electric power (P) of the

Journal of Physics: Conference Series **2793** (2024) 012006 doi:10.1088/1742-6596/2793/1/012006

blue laser. The 8-bit digital level of PWM ranging between 0 to 255 corresponding to the electric power (P) of the blue laser between 0 and 5 W, respectively. The temperature of the laser diode increases during the operation, so a small fan and an aluminum heatsink are used for cooling the laser diode. A transparent windshield is inserted beneath the laser head to protect the blowing hot air from the fan onto the opaque sheet. This ensures that the hot air does not affect the temperature of the under tested opaque sheet. As shown in figure 2(b), the laser illuminates entire the opaque material sheet and the highest light intensity is at around the center of the sheet. To measure the temperature (T) of the opaque sheet during the laser heating, a thermocouple wire is inserted between the sheet and the glass cuvette. In addition, the temperatures of the environment under (T_1) and upper (T_2) the transparent windshield are measured by two additional thermocouples.

3. Results and discussion

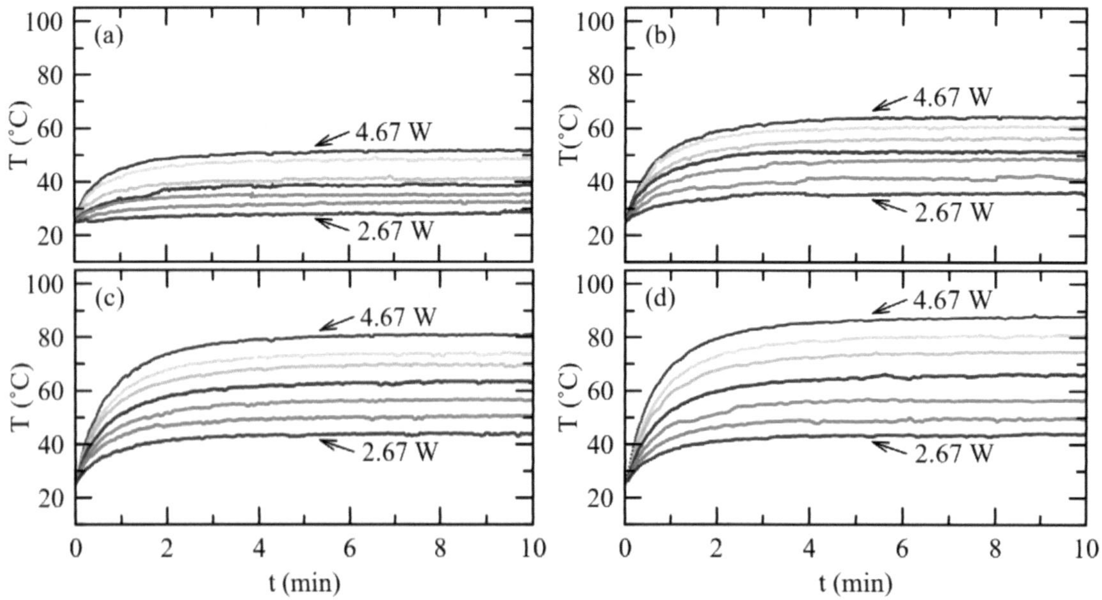

Figure 3. Temperature profile over time during laser heating for (a) silica bonded investment material, (b) gypsum bonded investment material, (c) silicone epoxy, and (d) mortar. From bottom to top curves, the input power P of laser are 2.67, 3.00, 3.33, 3.67, 4.00, 4.33, and 4.67 W, respectively.

Figure 3 shows changes of temperature (T) of opaque sheets made from different materials after the laser is switched on. The input electric power of the laser (P) is varied between 2.67 and 4.67 W (via the 8-bit digital level of PWM). For all experiments, the temperature increases rapidly in the first few minutes then it increases with a slow rate and reaches a maximum value T_{max} within around 10 minutes, except for silica bonded investment material with low power P of 2.67-3.33 W, the temperature increases slightly. It worths noting that during the experiment interval of 10 mins, the temperatures of the environment close to the under tested opaque sheet, i.e., under (T_1) and upper (T_2) transparent windshield increase from the room temperature 25°C to 27°C. Thus, the blowing hot air from the cooling fan does not play important role in the experimental results.

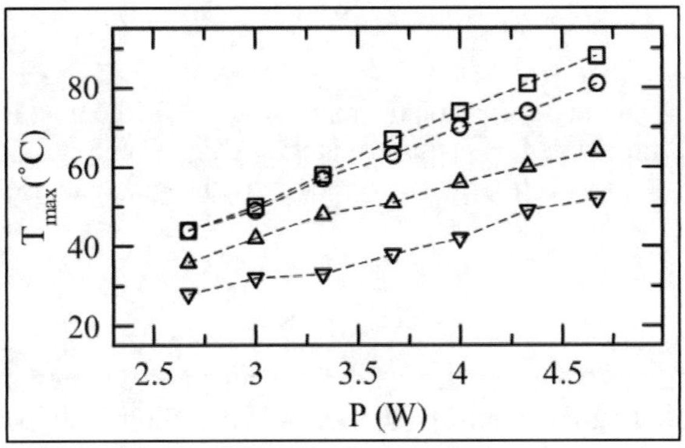

Figure 4. Maximum temperature T_{max} as function of the input power P of the laser. The symbols represent the materials as follows, □: mortar, O: silicon epoxy, Δ: gypsum bonded investment material, and ∇: silica bonded investment material.

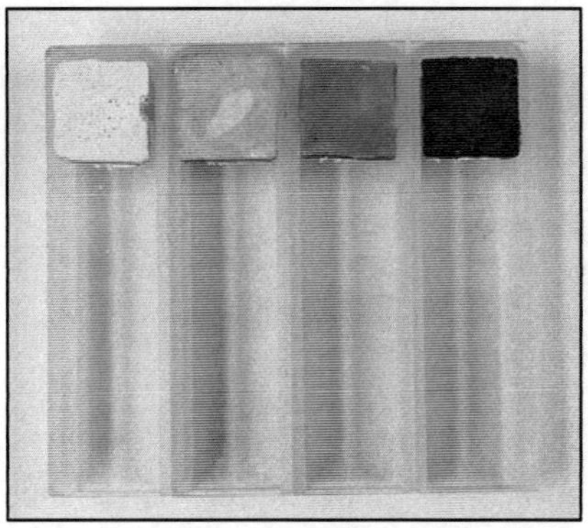

Figure 5. Opaque sheets on glass cuvettes after laser heating. From left to right, materials are a silica bonded investment material, a gypsum bonded investment material, a silicone epoxy, and a mortar.

A summarization of T_{max} as a function of the input power P is shown in figure 4. For all material types, T_{max} monotonously increases with P. For a given power P, T_{max} increases with the material type as the following, silica bonded investment material, gypsum bonded investment material, silicon epoxy, and mortar and at the maximum tested power of 4.67 W, their T_{max} are 52°C, 64°C, 81°C, and 88°C, respectively.

Figure 5 shows the opaque sheets after exposed by the laser beam several times (about 10 repetitions). For silica bonded investment material and gypsum bonded investment material sheets, their surfaces are lighter in color than those before the heating experiments (see figure 1) and some cracks are observed. In contrast, silicon epoxy and mortar sheets look very similar to those before laser heating. Based on the maximum temperature and the robustness against cracking, we recommend silicon epoxy or mortar to utilize as opaque material sheets for laser heating in atomic magnetometers.

We present the heating results with the electrical input power of the laser. The electrical-to-optical efficiencies of the blue laser diodes is typically about 15% to 48% at room temperature [16-17] and can be influenced by factors such as temperature of the diodes. Therefore, sufficient cooling for the laser diode as well as the driver circuit is recommended, especially during a long duration application of the laser heating.

This study demonstrates laser heating using four types of opaque flat sheets which conveniently prepared from powder. The presented methods can be adapted to prepare opaque sheets with other forms by using appropriate molds. For heating systems where the magnetic field generated by the opaque sheets is not an issue, pigments such as ferrous oxide powder which is widely used for modification of cement surfaces can be added to darkening the opaque sheets. In addition, the blue laser beam tends to damage some opaque sheets after reheating several times. For heating to a temperature higher than this study, we expect that a metal sheet may be a suitable opaque sheet to avoid such damage. It is worth noting that even though this investigation is motivated by atomic magnetometers, we expect that it is potentially useful for other fields where a compact setup for heating a small area is needed.

4. Conclusion

We have presented an investigation of heating small opaque sheets with dimensions 1×1×0.2 cm^3 using a blue laser diode from a low-cost engraving machine. The opaque material sheet is attached to a 1-cm glass cuvette mimicking a vapor cell in compact atomic magnetometers. A pulse width modulation is used to control the input electric power of the laser diode (P). For all experiments, the temperature of the opaque material sheet increases with time until it reaches a maximum temperature (T_{max}) within 10

Journal of Physics: Conference Series **2793** (2024) 012006 doi:10.1088/1742-6596/2793/1/012006

mins. T_{max} increases with the input power P of the laser and depends on the material type. At the highest P (4.67 W), silica bonded investment material, gypsum bonded investment material, silicone epoxy, and mortar have T_{max} about 52°C, 64°C, 81°C, and 88°C, respectively. Thus, mortar is the most suitable material for laser heating in atomic magnetometers.

Acknowledgments
This research has received funding support from the Department of Physics, Faculty of Science, Kasetsart University and the NSRF via the Program Management Unit for Human Resources & Institutional Development, Research and Innovation (grant number B05F650023), National Higher Education Science Research and Innovation Policy Council.

References

[1] Bekir S Y 2012 Introduction to Laser Heating *Laser Heating Applications* p 1-5
[2] Park C, Choi M, Lee S, Kim H, Lee T, Billah M M, Jung B and Jang J 2022 *Nanomaterials* 12 2127
[3] Pasang T, Lin P C, Misiolek W Z, Wei J Y, Masuno S, Tsukamoto M, Hori E, Sato Y, Tao Y, Yudhistiro D and Yunus S 2022 *Quantum Beam Sci.* **6** 24
[4] Tojo K, Masuno S, Higashino R and Tsukamoto M Works towards Goal of Next-Generation Laser Processing System with High-Power Blue Diode Lasers. Available online: https://www.industrial-lasers.com/cutting/article/16485172/japan-developsblue-laser-for-advanced-materials-processing (accessed on 17 September 2023)
[5] Jiménez-Martínez R and Knappe S 2017 Microfabricated Optically-Pumped Magnetometer *High Sensitivity Magnetometers Smart Sensors (Measurement and Instrumentation* vol 19) p 523–51
[6] Jundi L, Wei Q, Binquan Z, Zhuo W, Jixi L, Zhaohui H, Gang L and Jiancheng F 2018 *IEEE Sens. J.* **18** p 8198-8207
[7] Matthew J B, James L, Molly R, Ryan M H, Niall H, Elena B and Richard B 2022 *Trends in Nerosciences* **45** p 621-634
[8] Xuejing L, Weihong B, Yang L, Yanhui H, Min C and Xuedian Z 2022 *Microw. Opt. Technol. Lett.* **65** p 1516-1524
[9] Shah V, Knappe S, Schwindt P D D and Kitching J 2007 *Nature Photonics* **1** p 649-652
[10] Kun W, Binquan Z, Junjian T, Shaowen Z and Yaxiang W 2020 *SPIE* **11566**
[11] Savukov I and Romalis M V 2016 *Sensors* **16** 1690
[12] Mhaskar R, Knappe S and Kitching J 2012 *Appl. Phys. Lett.***101** 241105
[13] Sheng D, Perry A R, Krzyzewski SP, Geller S, Kitching J and Knappe S 2017 *Appl. Phys. Lett.***110** 031106
[14] Alem O, Mhaskar R, Ricardo J M, Sheng D,Leblance J, Trahms L, Sander T, Kitching J and Knappe S 2017 *Optics Exepress* **25** 7849
[15] Preusser J, Knappe S, Kitching J and Gerginov V 2009 *IEEE* p 1180-1182
[16] Muhammad A, Sven G, Bernhard S, Soenke T, Georg B, Andre S, Urs H, Christoph E, Matthias P, Alfred L, Andreas L, Harald K and Uwe S 2017 *IEEE High Power Diode Lasers and Systems Conference* p 47-48
[17] Shinichi N 2021 *Furukawa Electric Review* **52** p 10-14

Journal of Physics: Conference Series **2793** (2024) 012007 doi:10.1088/1742-6596/2793/1/012007

Ab initio study of gallium-doped zinc oxides and zinc-doped gallium oxides

Cheng-Lung Yu[1], Jian-Sheng Xie Su[2], Hsiang-Yu Hsieh[1], and Po-Liang Liu[*,1,3]

[1] Graduate Institute of Precision Engineering, National Chung Hsing University, Taiwan

[2] Taiwan Semiconductor Manufacturing Co., Ltd., Taiwan

[3] Department of Applied Materials and Optoelectronic Engineering, National Chi Nan University, Taiwan

*Corresponding author's e-mail: pliu@dragon.nchu.edu.tw

Abstract. This study utilizes *ab initio* methods to investigate the atomic structures of gallium-doped zinc oxides and zinc-doped gallium oxides, focusing on the mechanisms of $ZnGa_2O_4$ film formation through doping and substitution. Utilizing X-ray diffraction and bond length assessments, we investigate the interaction between zinc oxide and gallium oxide components. In the case of gallium doping within ZnO, the notable shift of the primary X-ray diffraction peak from 36.10° to 35.45° with an increase in gallium content closely aligns with the $ZnGa_2O_4$ peak at 35.50°. Conversely, when the substitution concentration of zinc in Ga_2O_3 exceeds 37.5%, it is not conducive to the formation of $ZnGa_2O_4$. This finding underscores the role of gallium doping in facilitating the development of $ZnGa_2O_4$ films, highlighting the nuanced impact of dopant concentration on the structural properties and demonstrating the preferential alignment of gallium-doped ZnO with the desired $ZnGa_2O_4$ structure.

1. Introduction

In contemporary electronics, electroluminescence (EL) technology is pivotal for a variety of applications, including backlights in automotive control panels [1], display technologies [2,3], and wearable electronics [4,5]. EL technology relies on phosphors that emit light under an electric field, with the emitted light's hue adjustable through the selection or doping of phosphor materials with specific metal ions [6-11]. Notably, D. Li's team synthesized blue-emitting zinc gallium oxide ($ZnGa_2O_4$) powder phosphors via a solid-state reaction [7], while P. Thiyagarajan's group achieved green emission from manganese-doped $ZnGa_2O_4$ films using pulsed laser deposition [10]. Phosphors are broadly classified into sulfide and oxide categories. Sulfide phosphors, as S. Tanaka highlighted, suffer from moisture sensitivity and reduced longevity, leading to instability [12]. In contrast, B. Qiao's research on chromium-doped $ZnGa_2O_4$ oxide phosphors produced via the citrate gel method demonstrated improved stability and performance in electroluminescent devices, underscoring the advantages of oxide phosphors over their sulfide counterparts [13]. These benefits include a direct correlation between luminance and applied voltage or frequency, underscoring oxide phosphors' superior stability and efficiency.

$ZnGa_2O_4$, an oxide phosphor, is distinguished by its exceptional chemical stability and low excitation voltage requirement. These attributes render it highly suitable for use in advanced display technologies, including thin film electroluminescent displays [14], field emission displays (FED) [15], and vacuum

Journal of Physics: Conference Series **2793** (2024) 012007 doi:10.1088/1742-6596/2793/1/012007

fluorescent displays (VFD) [16]. The compound crystallizes in the cubic system, characterized by a *Fd-3m* space group, and follows the AB_2O_4 molecular formula—where '*A*' is zinc (Zn) and '*B*' is gallium (Ga). Its spinel structure is defined by oxygen ions (O^{2-}) arranged in a face-centered cubic (FCC) lattice, with zinc ions (Zn^{2+}) and gallium ions (Ga^{3+}) occupying tetrahedral and octahedral sites, respectively. In essence, $ZnGa_2O_4$ is formed either by doping zinc oxide (ZnO) with gallium ions or gallium oxide (Ga_2O_3) with zinc ions. Research spearheaded by Y. S. Lee *et. al.* [17] focused on calculating formation energies to identify the most energetically favorable sites for Ga doping in ZnO. Formation energy is indicative of the energy necessary for crystal formation, with lower values suggesting easier formation. Their investigation into three Ga-doped ZnO models, i.e., substitution of Ga at Zn sites and doping at both octahedral and tetrahedral positions within ZnO, revealed that substitution at Zn sites resulted in the lowest formation energy, thus being the preferred method for Ga incorporation. In parallel, A. S. Goncalves's team [18] synthesized gallium-doped zinc oxide samples, examining their crystal structure via X-ray diffraction (XRD). These samples consistently displayed a cubic spinel crystal structure of $ZnGa_2O_4$, with a quantitative increase in $ZnGa_2O_4$ content correlating with higher doping levels, as determined by Rietveld's algorithm. Furthermore, C. Zhang and colleagues [19] undertook a theoretical study on the electronic properties of metal-doped β-Ga_2O_3. This analysis highlighted the presence of two distinct gallium sites, Ga_I and Ga_2, within the β-Ga_2O_3 structure. Prior works by C. Tang and W. Z. Xiao [20,21] established that substituting metal atoms at Ga_2 results in lower formation energy compared to substitutions at Ga_I. Notably, zinc's substitution into β-Ga_2O_3 was more energetically favorable at the Ga_I site, signifying zinc's appropriateness for doping at this location.

The utilization of $ZnGa_2O_4$ thin films within the domain of oxide phosphors is notably prevalent, underscoring the significance of understanding its synthesis mechanisms. Primarily, these mechanisms encompass the doping of zinc (Zn) into gallium oxide and the introduction of gallium (Ga) into zinc oxide. Remarkably, there exists a conspicuous absence of comprehensive comparative analyses between these two doping methodologies, particularly in terms of their impacts as doping concentrations are progressively augmented. This lacuna in the literature motivates our current investigation. Our research is designed to bridge this knowledge gap through a rigorous theoretical examination, wherein we will deploy computational models to simulate the doping process in both scenarios, replacing Zn with Ga in ZnO, and vice versa. The core of our analysis is to elucidate the resulting alterations in crystal structure, which we intend to achieve by meticulously analyzing lattice distortions, XRD patterns, and variations in bond lengths across an array of doping ratios. This approach aims not only to provide a granular understanding of the physical changes induced by doping but also to explore the underlying mechanisms governing these modifications. By delving into these comparative dynamics, our study aspires to offer novel insights into the synthesis of $ZnGa_2O_4$ thin films, potentially guiding future applications and enhancing the functional properties of oxide phosphors.

2. Computational methodology

In the present investigation, we engaged in theoretical calculations underpinned by first-principles methodologies, employing the Vienna *Ab initio* Simulation Package (VASP) to facilitate our computational analysis, and the PW91 approximation method within the Generalized Gradient Approximation (GGA) to ensure the precision and reliability of our simulations [22-25]. Our computational model for $ZnGa_2O_4$ comprises 56 atoms, including 8 zinc (Zn), 16 gallium (Ga), and 32 oxygen (O), conforming to the *Fd-3m* space group of its unit cell. To address computational efficiency while maintaining accuracy, we set the plane wave cut-off energy at 400 eV and employed a 5 × 5 × 5 Monkhorst-Pack grid for *k*-points sampling. In parallel, the zinc oxide (ZnO) model was developed utilizing a 2 × 2 × 2 supercell architecture, incorporating 32 atoms (16 Zn and 16 O) and adhering to the *P63mc* space group. The calculation parameters for ZnO mirrored those of $ZnGa_2O_4$, with a plane wave cut-off energy of 400 eV and a 4 × 4 × 2 gamma grid for *k*-points. For the gallium oxide (Ga_2O_3) model, a composition of 20 atoms (8 Ga and 12 O), respecting the *C2/m* space group, was employed. The cut-off energy was uniformly maintained at 400 eV across models, with a 2 × 8 × 4 gamma grid for *k*-points, ensuring consistency in our computational approach. Our exploration of doping mechanisms entailed the construction of models for gallium substitution in ZnO and zinc substitution in Ga_2O_3, meticulously optimized for geometric stability. Specifically, 15 models were devised for Ga-doping in ZnO, bifurcated into two distinct series based on the substitutional positioning within the ZnO matrix. The first series, designated $Ga_1Zn_{15}O_{16}$ and $Ga_xZn_{16-x}O_{16}$-1 (where x ranges from 2 to 8), features Ga atoms predominantly in the upper half of the ZnO supercell as shown in Figure 1. The second series, $Ga_xZn_{16-x}O_{16}$-2 (x = 2-8), distributes Ga atoms uniformly across the Zn lattice sites as shown in Figure 2. Geometric optimization of these models adhered to our standardized computational settings. Furthermore, our examination of Zn substitution in Ga_2O_3 yielded 14 models of $Zn_yGa_{8-y}O_{12}$-A_mB_n models, reflecting the strategic placement of Zn atoms in predefined positions within the Ga_2O_3 lattice, identified as sites A and B [19]. Note m and n are the number of atoms at A and B sites, respectively. These positions, informed by prior studies, underpin our systematic approach to doping shown in Figure 3. Each model underwent rigorous geometric optimization, aligning with the computational framework established for our entire dataset.

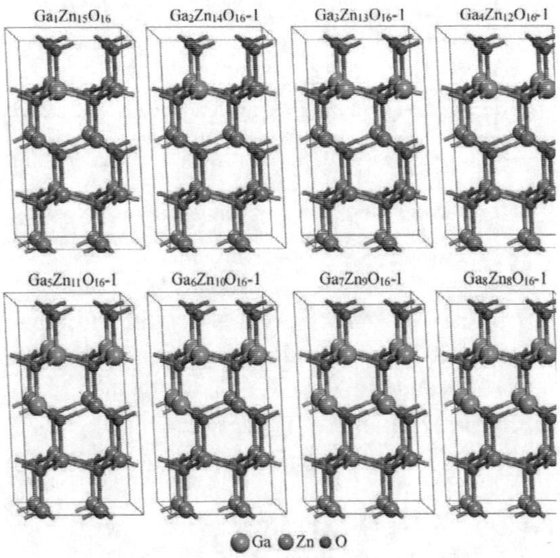

Figure 1. $Ga_1Zn_{15}O_{16}$ and $Ga_xZn_{16-x}O_{16}$-1 (x = 2~8) models are established in order to replace 2 to 8 Ga atoms in the upper half of the ZnO supercell. The atoms are represented by spheres: Ga (brown, large), Zn (gray middle size), and O (red, small).

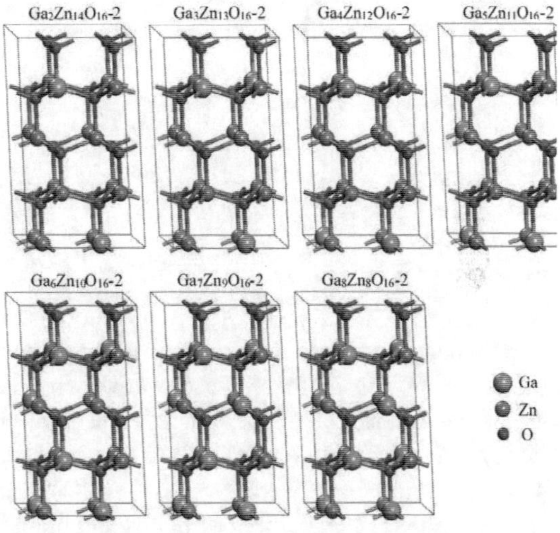

Figure 2. $Ga_xZn_{16-x}O_{16}$-2 (x = 2~8) models are established to uniformly replace 2 to 8 Ga atoms in the ZnO supercell. The atoms are represented by spheres: Ga (brown, large), Zn (gray middle size), and O (red, small).

3. Results and dsscussion

This study investigates the structural effects of Ga substitution and doping in ZnO, and Zn substitution and doping in Ga_2O_3, on the lattice dimensions, angles, and atom-to-atom bonding within the crystal lattice. Despite the distortions observed in lattice parameters and potential disruptions in atomic bonds, it is postulated that the overall connectivity within the lattice remains unchanged. For the Ga substitution in ZnO, we selected models with equivalent substitution levels, adhering to the principle that a lower total energy denotes a more stable lattice configuration. The equivalent substitution level here refers to the consistent proportion of gallium atoms replacing zinc atoms. Consequently, eight models with lower total energies were selected: $Ga_1Zn_{15}O_{16}$, $Ga_2Zn_{14}O_{16}$-2, $Ga_3Zn_{13}O_{16}$-2, $Ga_4Zn_{12}O_{16}$-2, $Ga_5Zn_{11}O_{16}$-2, $Ga_6Zn_{10}O_{16}$-2, $Ga_7Zn_9O_{16}$-2, and $Ga_8Zn_8O_{16}$-2. Similar criteria were applied to select four models for Zn substitution in Ga_2O_3, based on lattice stability indicated by lower total energy: $Zn_1Ga_7O_{12}$-A_1, $Zn_2Ga_6O_{12}$-A_2, $Zn_3Ga_5O_{12}$-A_3, and $Zn_4Ga_4O_{12}$-A_4. Bond analysis across 15 models listed in Table 1 reveals that the $Ga_7Zn_9O_{16}$-2 model is the closest structural resemblance to $ZnGa_2O_4$. This suggests a trend towards achieving the $ZnGa_2O_4$ crystal structure by increasing the doping concentration of Ga substitution in ZnO, highlighting the impact of substitution and doping on the crystal lattice integrity and stability. Bond length analyses for Ga-doped ZnO also reveal that Zn-O bond lengths closely approximate those in $ZnGa_2O_4$, remaining consistent across varying Ga concentrations from 6.3% to 44%. Conversely, Zn substitution into Ga_2O_3 results in Zn-O bond lengths that are considerably shorter than those in $ZnGa_2O_4$, indicating a preference for crystal structures other than $ZnGa_2O_4$ when Zn replaces Ga. XRD analysis was conducted on 15 models as shown in Figure 4. The undoped $Zn_{16}O_{16}$

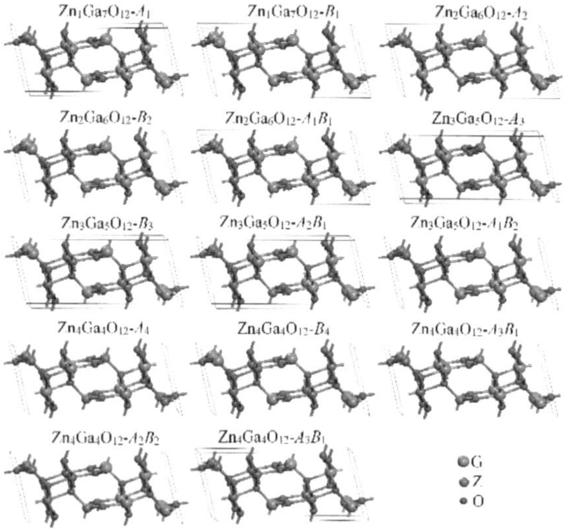

Figure 3. $Zn_yGa_{8-y}O_{12}$-A_mB_n (y = 1~8) models are established in order to replace 1 to 4 Zn atoms in the *A* or *B* sites in the β-Ga_2O_3 supercell. The atoms are represented by spheres: Ga (brown, large), Zn (gray middle size), and O (red, small).

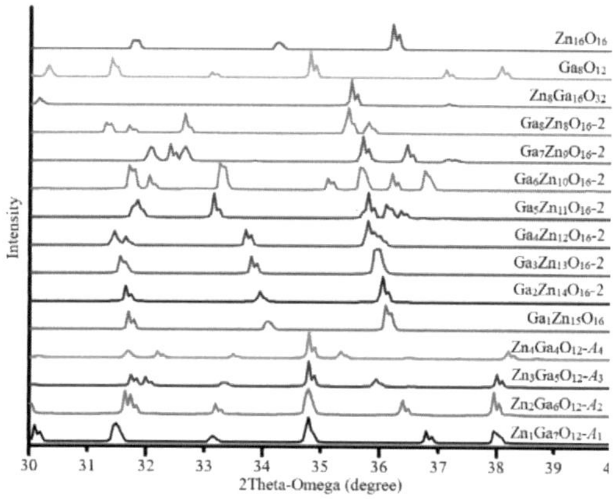

Figure 4. ZnO, β-Ga_2O_3, $ZnGa_2O_4$ and selected $Ga_1Zn_{15}O_{16}$, $Ga_xZn_{16-x}O_{16}$-1, $Ga_xZn_{16-x}O_{16}$-2, and $Zn_yGa_{8-y}O_{12}$-A_mB_n models, where the horizontal axis is the incident angle 2θ, and the vertical axis is the peak intensity.

Table 1. The number of bonds and bond lengths ZnO, β-Ga_2O_3, $ZnGa_2O_4$ and selected $Ga_1Zn_{15}O_{16}$, $Ga_xZn_{16-x}O_{16}$-1, $Ga_xZn_{16-x}O_{16}$-2, and $Zn_yGa_{8-y}O_{12}$-A_mB_n models.

Model	Bond Number		Bond Length			
	Zn-O	Ga-O	Zn-O	Strain (%)	Ga-O	Strain (%)
$Zn_{16}O_{16}$	64	-	1.98	-	-	-
Ga_8O_{12}	-	40	-	-	1.96	-
$Zn_8Ga_{16}O_{32}$	32	96	1.99	-	2.00	-
$Ga_1Zn_{15}O_{16}$	60	4	1.99	0	1.92	−4.00
$Ga_2Zn_{14}O_{16}$-2	56	8	2.00	0.50	1.93	−3.50
$Ga_3Zn_{13}O_{16}$-2	52	12	2.00	0.50	1.95	−2.50
$Ga_4Zn_{12}O_{16}$-2	48	16	2.01	1.01	1.97	−1.50
$Ga_5Zn_{11}O_{16}$-2	44	18	2.01	1.01	1.97	−1.50
$Ga_6Zn_{10}O_{16}$-2	40	21	2.01	1.01	1.98	−1.00
$Ga_7Zn_9O_{16}$-2	36	26	2.00	0.50	2.00	0
$Ga_8Zn_8O_{16}$-2	32	32	2.04	2.51	2.01	0.50
$Zn_1Ga_7O_{12}$-A_1	4	36	1.92	−3.52	1.96	−2.00
$Zn_2Ga_6O_{12}$-A_2	8	32	1.91	−4.02	1.97	−1.50
$Zn_3Ga_5O_{12}$-A_3	12	28	1.92	−3.52	1.98	−1.00
$Zn_4Ga_4O_{12}$-A_4	16	24	1.92	−3.52	2.01	0.50

unit cell exhibited characteristic peaks at 2θ values of 31.75°, 34.25°, and 36.20°, while the Ga_8O_{12} unit cell displayed peaks at 2θ values ranging from 30.30° to 38.05°. The $Zn_8Ga_{16}O_{32}$ unit cell showed characteristic peaks at 30.15°, 35.50°, and 37.15°. The doping of Ga into ZnO resulted in a gradual peak shift from 36.10° to 35.45° with increasing the gallium concentration from 6.3% to 50% in XRD patterns, suggesting a structural transition towards $ZnGa_2O_4$ characteristics, especially evident in the $Ga_7Zn_9O_{16}$-2 model which close to the characteristic peak of $ZnGa_2O_4$ at 35.50°. Conversely, in gallium oxide, substituting gallium atoms with zinc and increasing the zinc concentration from 12.5% to 50% results in the observation of a shift in the X-ray diffraction peaks from 36.80° to 35.30°. Beyond a 37.5% concentration of zinc, the diffraction peaks deviate away the characteristic peak of $ZnGa_2O_4$ at 35.50°, indicating that the phase transition of the compound structure is unfavorable to the formation of $ZnGa_2O_4$.

4. Conclusions

This study applies *ab initio* density functional theory to explore the synthesis of $ZnGa_2O_4$ thin films through the atomic structure modifications induced by gallium doping in zinc oxide and zinc doping in gallium oxide. By prioritizing models based on their stability (inferred from low total energy), the research advances our understanding of substitution doping's impact on material properties. The investigation reveals that gallium substitution in zinc oxide progressively aligns the crystal structure closer to that of $ZnGa_2O_4$, as evidenced by X-ray diffraction patterns and bond length consistency. Notably, the $Ga_7Zn_9O_{16}$-2 model emerges as particularly significant, closely emulating the characteristic structure of $ZnGa_2O_4$, suggesting that gallium substitution in zinc oxide is a more effective pathway for synthesizing $ZnGa_2O_4$ thin films. Conversely, zinc substitution in gallium oxide deviates from this desired structure. This comprehensive analysis underscores the promising avenue of gallium substitution in zinc oxide for achieving $ZnGa_2O_4$, thin films, highlighting its implications for enhancing materials science and thin film technology.

References

[1] Rack P D and Holloway P H 1998 The structure, device physics and material properties of thin film electroluminescent displays Materials Science and Engineering: R: Reports vol 21 pp 171-219

[2] Li S, Peele B N, Larson C M, Zhao H and Shepherd R F 2016 A stretchable multicolor display and touch interface using photopatterning and transfer printing Advanced Materials vol 28 pp 9770-75

[3] Stauffer F and Tybrandt K 2016 Bright stretchable alternating current electroluminescent displays based on high permittivity composites Advanced Materials vol 28 pp 7200-03

[4] Liang G, Yi M, Hu H, Ding K, Wang L, Zeng H, Tang J, Liao L, Nan C, He Y and Ye C. 2017 Coaxial-structured weavable and wearable electroluminescent fibers Advanced Electronic Materials vol. 3 p 1700401

[5] Zhang Z, Guo K, Li Y, Li X, Guan G, Li H, Luo Y, Zhao F, Zhang Q, Wei B, Pei Q and Peng H 2015 A colour-tunable, weavable fibre-shaped polymer light-emitting electrochemical cell Nature Photonics vol 9 pp 233-38

[6] Kim J S, Kang H I, Kim W N, Kim J I, Choi J C, Park H L, Kim G C, Kim T W, Hwang Y H, Mho S I, Jung M C and Han M 2003 Color variation of $ZnGa_2O_4$ phosphor by reduction-oxidation processes Applied Physics Letters vol 82 pp 2029-31

[7] Li D, Wang Y, Xu K, Li L and Hu Z 2015 Persistently luminescent and photocatalytic properties of $ZnGa_2O_4$ phosphors Materials Research Express vol 2 p 046202

[8] Yang S H, Hsueh T J, and Chang S J 2007 Effect of Substrate Properties on Luminescence of White $ZnGa_2O_4$ Phosphor Japanese Journal of Applied Physics vol 46 pp 4166-69

[9] Kim J S, Lee S G, Park H L, Park J Y and Han S D 2004 Optical and electrical properties of $ZnGa_2O_4/Mn^{2+}$ powder electroluminescent device Materials Letters vol 58 pp 1354-57

[10] Thiyagarajan P, Kottaisamy M, Rama N and Rao M S R 2009 Study of pulsed laser deposited $ZnGa_2O_4$: Mn phosphor thin films in an oxygen controlled environment Journal of Physics D Applied Physics vol 42 p 155301

[11] Luan T, Liu J, Yuan X and Li J G 2017 Controlled Hydrothermal Synthesis and Photoluminescence of Nanocrystalline $ZnGa_2O_4$: Cr^{3+} Monospheres Nanoscale Research Letters vol 12 p 219

[12] S. Tanaka 1990 Thin-film electroluminescent devices using CaS and SrS Journal of Crystal Growth vol 101 pp 958-66
[13] Qiao B, Tang Z L, Zhang Z T and L. Chen 2007 Study on $ZnGa_2O_4$: Cr^{3+} a.c. powder electroluminescent device Materials Letters vol 61 pp 401-04
[14] Minami T, Kuroi Y and Takata S 1996 Preparation of $ZnGa_2O_4$: Mn phosphor thin films as emitting layers for electroluminescent devices Journal of Vacuum Science & Technology A: Vacuum, Surfaces, and Films vol 14 pp 1736-40
[15] Yu M, Lin J, Zhou Y H and Wang S B 2002 Citrate-gel synthesis and luminescent properties of $ZnGa_2O_4$ doped with Mn^{2+} and Eu^{3+} Materials Letters vol 56 pp 1007-13
[16] Shea L E, Datta R K and Brown J J 1994 Low Voltage Cathodoluminescence of Mn^{2+}-Activated $ZnGa_2O_4$ Journal of The Electrochemical Society vol 141 pp 2198-200
[17] Lee Y S, Peng Y C, Lu J H, Zhu Y R and Wu H C 2014 Electronic and optical properties of Ga-doped ZnO Thin Solid Films vol 570 pp 464-70
[18] Goncalves A S, Lima S A M, Davolos M R, Antonio S G, Santos C O P 2006 The effects of $ZnGa_2O_4$ formation on structural and optical properties of ZnO: Ga powders Journal of Solid State Chemistry vol 179 pp 1330-34
[19] Zhang C, Liao F, Liang X, Gong H, Liu Q, Li L, Qin X F, Huang X and Huang C J 2019 Electronic transport properties in metal doped beta-Ga_2O_3: A first principles study Physica B Condensed Matter vol 562 pp 124-130
[20] Tang C, Sun J, Lin N, Jia Z, Mu W, Tao X and Zhao X 2016 Electronic structure and optical property of metal-doped Ga_2O_3: a first principles study RSC Advances vol 6 pp 78322-34
[21] Xiao W Z, Wang L L, Xu L, Wan Q and Zou B S 2009 Electronic structure and magnetic interactions in Ni-doped β-Ga_2O_3 from first-principles calculations Scripta Materialia vol 61 pp 477-80
[22] Kresse G and Hafner J 1994 Norm-conserving and ultrasoft pseudopotentials for first-row and transition elements Journal of Physics: Condensed Matter vol 6 pp 8245-57
[23] Kresse G and Furthmüller J 1996 Efficiency of ab-initio total energy calculations for metals and semiconductors using a plane-wave basis set Computational Materials Science vol 6 pp 15-50
[24] Kresse G and Furthmüller J 1996 Efficient iterative schemes for ab initio total-energy calculations using a plane-wave basis set Physical Review B vol 54 pp 11169-86
[25] Kresse G and Joubert D 1999 From ultrasoft pseudopotentials to the projector augmented-wave method Physical Revi vol 59 pp 1758-75

Predicted yield and detectability of GW events from the IllustrisTNG simulations

Luchang Niu[1,*], Tiangang Zhang[2], Kun Zhang[2] and Yiting Li [3]

[1]Department of Physics and Astronomy, University of Rochester, 14627, USA
[2]School of Physics, Beihang University, Beijing, 102600, China
[3]University of Michigan, Ann Arbor, department of astronomy, USA 48109

* Corresponding author's email: niuluchang12345@126.com

Abstract. The forthcoming generation of space-based gravitational wave (GW) observatories promises robust capabilities for multi-bandwidth measurements, particularly for the study of supermassive black holes (SMBHs). In this research, we employ the IllustrisTNG simulation to investigate GW signals originating from SMBHs. Our approach involves extracting the merger history of black holes (BHs) from the simulation data and utilizing a phenomenological waveform model to calculate the GW strains, signal-to-noise ratios and event rates. Notably, detectors like LISA, Taiji, and Tianqin demonstrate the potential to detect SMBHs with masses exceeding approximately $10^6 M_\odot$, and exhibit the capability to detect one merger in the Merger-Ringdown phase and one merger in the Inspiral phase annually, at redshifts reaching up to $z \sim 6$.

1. Introduction

Gravitational waves (GWs), first envisioned by Albert Einstein in his 1916 general theory of relativity, are spacetime ripples generated by massive accelerating objects[1], which can be detected by GW detectors. While ground-based detectors cannot detect sub-1 Hz gravitational waves necessary for studying massive black holes (MBHs) and supermassive black holes (SMBHs) due to terrestrial gravity gradient noise. However, these lower-frequency waves are observable with space-based instruments

Space-based interferometers, such as the Laser Interferometer Space Antenna[2], operate within a frequency range of 10^{-4} to 10^{-1} Hz. This range enables the detection of mergers involving massive black holes (MBHs) in the mass spectrum of approximately 10^4 to $10^7 M_\odot$during the reionization epoch[3]. Research groups employ diverse methods, including analytical calculations[4] and cosmological simulations[5], to estimate black hole (BH) merger rates and predict gravitational wave (GW) signals.

2. Methods

2.1. Cosmological simulations

The IllustrisTNG project[6] comprises a collection of advanced cosmological simulations that model the formation of the universe. These simulations encompass a significant portion of a simulated universe, starting from the early stages after the Big Bang and progressing to the present day. They incorporate various physical processes that influence galaxy formation. There are three volumes in the original

IllustrisTNG project with a total of 18 simulations, each demonstrating variations in physical size, mass resolution, etc. [6]. The side length of simulation box is 35000.0, 75000.0, 205000.0 [ckpc/h] for TNG50, TNG100 and TNG300 respectively. TNG100 are used in this study. The simulations use the cosmological parameters of Planck2015 (i.e. total matter density $\Omega_m = 0.3089$, dark energy density $\Omega_\Lambda = 0.6911$, baryonic matter density $\Omega_b = 0.0486$, hubble constant 'little h' $h_0 = 0.6774$ [100km/s/Mpc]). The number of dark matter particles of TNG100 is 6028568000 with no gas tracer particles.

2.1.1. *Black Hole Mergers.* Merging binary black hole (BBH) systems are characterized by intrinsic parameters: individual BH masses (m_1 and m_2), spin angular momenta ($\overrightarrow{S_1}$ and $\overrightarrow{S_2}$), leading to dimensionless spin magnitudes $\chi_{1,2} \equiv c|\overrightarrow{S_{1,2}}|/(Gm_{1,2}^2)$, and the gravitational wave frequency f. Additionally, luminosity distance computed from redshift and orientation angles (ψ, ι, α, δ) are needed. The gravitational waveform produced by a binary black hole (BBH) merger depends on the system's mass ratio, spin, and distance.

The BH merger process comprises three phases: the inspiral phase, the merger phase and the ringdown phase. The inspiral phase is driven by gravitational radiation damping, causes a gradual orbital decay. It ends when the black hole separation approaches the innermost stable circular orbit (ISCO) with $r_{isco} > 6GM/c^2$, where $M = m_1 + m_2$. During this phase, binary dynamics and gravitational waveforms are described using post-Newtonian expansions[7]. The merger phase begins when the binary reaches its highest orbital frequency (ISCO or merger frequency f_{isco}), or when the radiation-reaction timescale matches the orbital period. The separation decreases rapidly, transitioning from inspiral to plunge, occurring for orbits more tightly bound than the ISCO[8]. In the ringdown phase, the system settles into a Kerr black hole state, emitting gravitational waves due to deviations from its final axisymmetric form. This phase's dynamics are accurately described as oscillations of the final black hole's quasinormal modes, using perturbation theory on the Kerr spacetime background[9].

2.1.2. Treatment of the Inspiral and Merger Phases for the BBHs from the Cosmological Simulations. In cosmological simulations, the formation of a binary black hole (BBH) occurs when two black holes become gravitationally bound. To model the inspiral and merger phases, we adopt a similar approach as employed in other studies[5]. We initiate the inspiral phase when the BBH enters the spatial resolution range, and the timescale for inspiral is calculated as follows.

In the Newtonian limit, the orbital radius r and orbital angular velocity Ω are interconnected through the Keplerian relationship (Unless specified otherwise, we use the geometric system of units in which $G = c = 1$): $\frac{M}{r} \equiv \frac{1}{\tilde{r}} \equiv (M\Omega)^{2/3} \equiv \tilde{\Omega}^{2/3}$. Here, $\tilde{r}$ and $\tilde{\Omega}$ represent the dimensionless radius of the orbit and the dimensionless orbital angular velocity, respectively. The gravitational wave frequency for a specific harmonic mode m is expressed as $f_m = \frac{m}{2\pi M\tilde{\Omega}}$. Our analysis focuses only on the dominant $m = 2$ mode, for which the associated frequency is given by $f_2 = \frac{\Omega}{\pi}$.

In the limit where $m_2 \gg m_1$, the total duration of the inspiral phase until the innermost stable circular orbit (ISCO) is attained[10], denoted as T_{insp}, is given by:

$$T_{\rm insp} = \frac{5}{256}\frac{1}{\eta}\frac{M}{\tilde{\Omega}^{8/3}}\tau = 1.41 \times 10^6 \text{ sec}\left(\frac{f_2}{10 \text{ mHz}}\right)^{-8/3}\left(\frac{10}{m_1}\right)\left(\frac{10^6 M_\odot}{M}\right)^{2/3}\tau$$
$$= 0.141 \text{ sec}\left(\frac{f_2}{100 \text{ Hz}}\right)^{-8/3}\left(\frac{10M_\odot}{m_1}\right)\left(\frac{100M_\odot}{M}\right)^{2/3} \quad (1)$$

Here, $\eta = m_1 m_2/M^2$ represents the symmetric mass ratio, with M being the total mass. Additionally, τ denotes the relativistic correction to T_{insp}, a function dependent on the orbital radius r/r_{isco} and the black hole spin parameter[11]. The parameter r/r_{isco} can be determined as the ratio of f_2 and f_{isco}

Journal of Physics: Conference Series **2793** (2024) 012008 doi:10.1088/1742-6596/2793/1/012008

concurrently: $\frac{r}{r_{isco}} = (f/f_{isco})^{-2/3}$. The ISCO is located at $\tilde{r}_{isco}$, where the total energy $E(\tilde{r})$ of the object is minimized, or equivalently, where the time derivative of Ω becomes infinite. The value of $\tilde{r}_{isco}$ can be determined as the root of the quartic equation $\tilde{r}^2 - 6\tilde{r} + 8\chi\tilde{r}^{1/2} - 3\chi^2 = 0$. This value lies between 1 (when χ =1) and 6 (when χ =0). In our analysis, where the BH spin is assumed to be 0, the frequency at the ISCO is calculated as:

$$f_{2,\text{isco}} = \frac{\Omega}{\pi} = \frac{\tilde{\Omega}_{isco}}{\pi M} = \left(\frac{1}{\tilde{r}_{isco}}\right)^{3/2} / \pi M = \frac{\tilde{r}_{isco}{}^{-3/2}}{M\pi} = \frac{6^{-3/2}}{M\pi} \quad (2)$$

For each binary black hole (BBH) obtained from the cosmological simulations, we incorporate the timescale T_{insp} to represent the inspiral phase. Beyond this period, we assume that the BBH coalesces, transitioning into the merger phase.

2.2. Gravitational wave calculations

In our calculations of gravitational waves emanating from binary black holes (BBHs), we utilize the notation Ξ to represent the physical parameters of a BH binary, where $\Xi \in (M, \eta, \chi_1, \chi_2)$. Here, M is the total mass of the BBH, η signifies the mass ratio, and χ_1 and χ_2 are the spin parameters of the BHs. Here, the spin parameters χ_i fall within the range of $[-1,1]$.

We utilize the "PhenomD" model[12] in our study to compute GW amplitudes in the Fourier domain. The "PhenomD" model is a hybrid waveform model that combines an analytical effective-one-body (EOB) model with numerical relativity (NR) simulation data. Within this model, the Inspiral phase is described using the uncalibrated EOB model from the minimum frequency up to the transition frequency f_{isco}. Beyond this frequency, the model switches to NR simulations to accurately model the Merger-Ringdown waveforms spins[13]. The amplitude of the gravitational wave signal from such a system also depends on the luminosity distance[14].

The sensitivity of a GW detector is often described by the root power spectral density (PSD), $\sqrt{S(f)}$. In contrast, the amplitude of a GW source can be expressed by the characteristic strain. The square root PSD that describes the strain noise[15] is given by:

$$\sqrt{S_n(f)} = h_n(f) f^{-\frac{1}{2}} \quad (3)$$

Where the detector noise curve $S_n(f)$, may be acquired through either analytical fitting procedures or interpolation techniques applied to the sensitivity data. For this study, we select five space-borne interferometers Tianqin, Taiji, LISA, BBO, and DECIGO. In table 1, we list the GW detectors considered in this study, and summarize the analytical approximations of the PSD for each detector. For LISA, Taiji and Tianqin, we adopt the analytical fits[6,34]. The analytical approximations of the BBO and DECIGO noise curves are given by [17].

3. Results

3.1. Gravitational wave characteristic strain from black hole mergers

When a GW passes a detector, it imparts a fractional alteration in the detector's arm lengths. This alteration is captured and documented as a dimensionless strain with respect to time, given by $\text{h(t)} = \frac{\Delta \text{L(t)}}{\text{L}}$. The characteristic strain is designed to incorporate the effect of integrating an inspiralling signal[15], whose amplitude is given by:

$$h_c(f) = 4f^2 \left|\tilde{h}(f)\right|^2 \quad (4)$$

The symbol f represents the rest-frame frequency of the source, and $\tilde{h}(f)$ corresponds to the Fourier transform of the strain signal, which can be expressed as $\tilde{h}(f) = \mathcal{F}\{h(t)\}(f) = \int_{-\infty}^{\infty} h(t) e^{-2\pi i f t}$. Likewise, the characteristic noise strain pertaining to GW detectors can be calculated through a rearrangement of equation (3) where the term $S_n(f)$ signifies the analytical approximation of the noise spectral densities, as itemized in table 1.

Journal of Physics: Conference Series **2793** (2024) 012008 doi:10.1088/1742-6596/2793/1/012008

We adhere to a method[5] to compute $h_c(f)$ and $h_n(f)$ from our simulation data, which involves the following steps:

- Extract the merger tree of the simulation and determine the intrinsic parameters of each merging BBH pair at the moment of "merger" at a specific redshift (i.e., the merger redshift)..
- Segregate the coalescence process into the inspiral and merger-ringdown phases, in accordance with the method in section 2.1.2. The inspiral phase starts when the distance between the BBH pair falls below the spatial resolution threshold and continues for a duration of $T_{\rm insp}$ as defined in equation (1). Subsequently, the pair coalesces and enters the merger-ringdown phase.
- Utilize equation (2)) to compute the frequency at the moment of merger (the merger frequency $f_{\rm isco}$). Convert this frequency to the observed frequency $f_{\rm obs}$, via the relation $f_{\rm obs} = f_{\rm isco}/(1+z)$.
- Calculate the frequency-domain amplitudes of the GW signals $|\tilde{h}(f_{\rm obs})|$ with the "PhenomD" model, for all BH coalescence events in TNG100. This calculation assumes no BH spin.
- Use equation (4) to compute the characteristic strain amplitude at the observed merger frequency $h_c(f_{\rm obs})$.
- Use equation (5) to calculate the characteristic noise amplitude at the observed merger frequency $h_n(f_{\rm obs})$ of each GW detector.
- Utilize the "PhenomD" model to compute the theoretical projections of $h_c(f)$ for equal-mass, non-spinning BBH systems across a mass range spanning from $M_1 = M_2 = 10^4 M_\odot$ to $M_1 = M_2 = 10^{10} M_\odot$), at various redshifts, ranging from $z \approx 15$ to $z \approx 10^{-3}$.

Table 1. Parameters description of the corner plot.

Detector	Arm length (km)	Frequency (Hz)	$S_n(f)$
LISA	5×10^6	$3\times10^{-5}-1$	$\frac{10}{3L^2}\left(P_{dp}+2\left(1+cos^2\left(\frac{f}{f_*}\right)\frac{P_{acc}}{(2\pi f)^4}\right)\right)\left(1+0.6\left(\frac{f}{f_*}\right)^2\right)$
BBO	5×10^4	$10^{-3}-10$	$2\times10^{-49}f^2+4.58\times10^{-49}+1.26\times10^{-52}f^{-4}$
TIANQIN	$\sqrt{3}\times10^5$	$10^{-4}-1$	$\frac{10}{3L^2}\left[10^{-24}+\frac{4\times10^{-30}}{(2\pi f)^4}\left(1+\frac{10^{-4}Hz}{f}\right)\right]\times\left[1+0.6\left(\frac{f}{f_*}\right)^2\right]$
TAIJI	3×10^6	$10^{-4}-1$	$\frac{10}{3L^2}\left(P_{dp}+2\left(1+cos^2\left(\frac{f}{f_*}\right)\frac{P_{acc}}{(2\pi f)^4}\right)\right)\left(1+0.6\left(\frac{f}{f_*}\right)^2\right)$
DECIGO	10^3	$0.1-10$	$7.05\times10^{-48}[1+(f/7.36)^2]+4.8\times10^{-51}f^{-4}\frac{1}{1+(f/7.36)^2}+5.33\times10^{-52}f^{-4}$

[A] THE APPROXIMATED ANALYTICAL NOISE SPECTRAL DENSITY IN EITHER $S_N(f)$ OR $S_N(x)$, WHERE x IS DIMENSIONLESS FREQUENCY $x = f/f_0$. $f_* = \frac{c}{2\pi L}$. FOR LISA, $P_{dp} = (15\times10^{-12}m)^2\left(1+\left(\frac{2mHz}{f}\right)^4\right)Hz^{-1}$, $P_{acc} = (3\times10^{-15}ms^{-2})^2\left(1+\left(\frac{0.4mHz}{f}\right)^2\right)\left(1+\left(\frac{f}{8mHz}\right)^4\right)Hz^{-1}$. FOR TAIJI, $P_{dp} = (8\times10^{-12}m)^2\left(1+\left(\frac{2mHz}{f}\right)^4\right)Hz^{-1}$, S_{acc} IS SAME AS LISA.

Figure 1 displays the resultant GW characteristic strain amplitudes $h_c(f_{\rm obs})$, for all SMBH mergers at various redshifts in TNG100. These are compared with theoretical forecasts for equal-mass mergers at different redshifts. Additionally, sensitivity curves for GW detectors, namely Tianqin, LISA, Taiji, DECIGO, and BBO, are featured for reference. It's important to note that figure 1 only presents the $h_c(f_{\rm obs})$ values corresponding to the Merger-Ringdown phase for illustrative purposes.

Figure 1 shows a trend that the characteristic strain amplitude increases as the masses of the BH mergers increase, with decreased observed frequency. For these massive BH mergers of $\gtrsim 10^6$– $10^9\ M_\odot$, BBO and DECIGO can detect part of BH mergers of $10^6\ M_\odot$ at very low redshift dowm to $z\ \sim 1$, while LISA can detect mergers of ∽ 10^6– $10^{7.5}\ M_\odot$ in the redshift range $z \sim 7 - 10^{-3}$, but for more massive mergers of $\geq 10^8\ M_\odot$ and above, they move out of LISA's frequency range. For Tianqin and Taiji, they have a similar sensitivity curves as LISA but with a higher cut-off frequency which makes them detect less massive BH mergers than LISA especially these BHs of $\gtrsim 10^7 - 10^8\ M_\odot$, but they can still detect mergers in the same range of redshift as LISA. The gravitational wave signals produced by the SMBH mergers shown in figure 1 closely agree with the results of other study[5].

3.2. Detectability of black hole mergers

A GW detector generates an output signal that combines noise, characterized by the noise PSD $S_n(f)$, and a potential GW signal: $s(t) = n(t) + h(t)$. The separation of contributions from the signal and the noise can be accomplished through the convolution of the filter function associated with the Wiener filter with the detector's output[15].

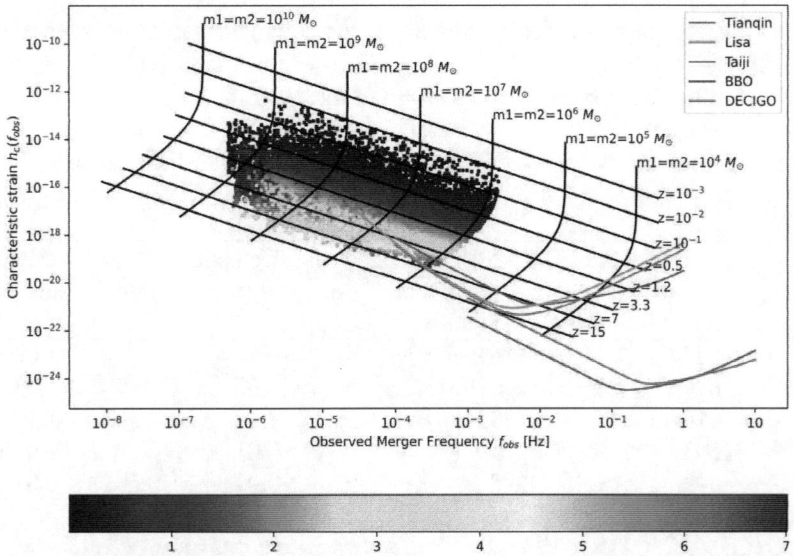

Figure 1. Colored data points represent BBH merger events in TNG100. These points are color-coded based on the redshift at the time of merger. Additionally, sensitivity curves for space-based gravitational wave detectors are displayed. Alongside them, black gridlines depict theoretical predictions for h_c originating from equal-mass, non-spinning BBH mergers. Data points positioned above a detector's sensitivity curve may be detectable by that particular detector, while those below the curve represent merger events that cannot be resolved by the detector.

we employ the matched-filtering methodology and the root-mean-square value for the SNR $\rho_{\rm rms}$ can be expressed as: $\rho_{\rm rms}^2 = 4\int_0^\infty \frac{|\tilde{h}(f)|^2}{S_n(f)} df$, where $\tilde{h}(f) = \int_{-\infty}^{\infty} e^{2\pi i f t} h(t)\, dt$ is the Fourier transform of the dimensionless strain $h(t)$ generated from the ``PhenomD" model.In the case of an inspiraling source, equation above no longer remains applicable. The precise definition of the characteristic strain h_c, is derived directly from the definition of SNR as $\rho_{\rm rms} = \sqrt{\int \left(\frac{h_{c,m}(f_m)}{h_n(f_m)}\right)^2 d\ln(f_m)}$ [18]. In the Merger-Ringdown phase, we establish the upper integration limit as infinity and the lower limit $f_{\rm isco}$ as expressed in Equation (2). The square of the SNR for the Merger-Ringdown phase can be precisely expressed as:$\rho^2 = 4\int_{f_{\rm isco}}^\infty \frac{|\tilde{h}(f)|^2}{S_n(f_{\rm obs})} df_{\rm obs}$.

3.3. Event rate of black hole mergers

The count of detectable sources surpassing the detection threshold $SNR = 10$ within a redshift range from z to $z + dz$ is defined as $\bar{N}(z,{\rm SNR} \geq 10)/\Delta z V_c \sim d^2\bar{n}(z,{\rm SNR} \geq 10)/dz dV_c$, where V_c represents the co-moving volume. The estimated event rate of detected GW sources per observed time, considering the integration across all redshifts[19], is provided by:

$$\frac{d\bar{N}}{dt_{\text{obs}}} = \int_0^{\infty} \frac{d^2\bar{n}(z, SNR \geq 10)}{dz} \frac{dz}{dt} \frac{dV_c}{dz} \frac{dz}{(1+z)} \tag{5}$$

To assess the count of resolvable events originating from inspiral and merger-ringdown sources within the binary black hole (BBH) simulations, we designate signals with SNR ≥ 10 whose observed merger frequency falls within the frequency range of the target detector (as specified in table 1) as detectable. Because TNG100 lack the resolution required to determine the spin of each individual BH, we employ $\chi_1 = 0$ and $\chi_2 = 0$ for the merging BH pairs. Recent research[5] suggests that spin contributions to the event rate across all phases are negligible.

Table 2 provides an overview of the approximated counts of detectable signals and the corresponding event rates originating from inspiral, merger-ringdown, and complete inspiral-merger-ringdown sources within TNG100, considering various GW detectors. For the super massive black holes ($M_{\odot} \geq 10^6$), only very small part of the BBH events in TNG100 fall inside of the frequency range of the BBO and DECIGO, so they will be able to resolve less BH mergers than other detectors.

Table 2. Estimates of resolvable signals (SNR ≥ 10) and event rates of inspiral, merger-ringdown (Mrd), and full inspiral-merger-ringdown (full-IMR) sources from IllustrisTNG for various gravitational wave detectors.

Detector	Number of Resolvable Inspiral Signals	Number of Resolvable Mrd Signals	Detection Rates of Inspiral Signals (yr^{-1})	Detection Rates of Mrd Signals (yr^{-1})
Tianqin	6757	6773	1.00	1.02
Taiji	6796	6796	1.03	1.03
LISA	9167	9248	1.18	1.19
DECIGO	593	593	0.01	0.01
BBO	593	593	0.01	0.01

[a] Numbers of mergers events is 18282.

The number of resolvavle signals and event rate of BBO and DECIGO are same because their sensitivity curves are very similiar and do not reach to the area of bh mergers(see figure 1). Each of Tianqin, Taiji, LISA will be able detect about 1 merger event of Merger-Ringdown phase and Inspiral phase per year respectively while BBO and DECIGO will not be able to detect almost any signals of either phase per year.

3.4. Parameters distribution of detectable events

The distribution of different parameters of SMBH merger events is an important indicator of the environments in which they grow, the star formation rate (SFR). They are very useful when study the origin and formation of these super massive black holes.

Figure 2 shows the corner plot super massive bh mergers in IllustrisTNG. The hists in green are the parameters (Table 3) distributions of resolvable mergers events while the hists in blue are distributions of irresolvable merger events. The red contours describe the relationship between these parameters related resolvable merger events.

We can see from figure 2 that the distribution of Rho(Local gas density in the vicinity of the blackhole) is the superposition of two normal distribution with two differnet peaks. However the BH mergers resolvable by LISA are only distributed by one of the normal distribution. Almost all the BHs of another normal distribution are irresolvable. Similarity, Mdot(The mass accretion rate onto the black hole, instantaneous) has two peaks as well but GW detectors can only detect BHs gathering around one of the peaks. In Rho-Mdot plot, there are clearly three peaks but only BHs around one of the peaks can be detected by LISA. In fact, we have a relation between Cs and Rho:$\dot{M_{BH}} = (4\pi G^2 M_{BH}^2 \rho)/(c_s^2 + v_{rel}^2)^{\frac{3}{2}}$, where ρ and c_s are the Local gas density and Local gas sound-speed in the vicinity of the black hole and v_{rel}

Journal of Physics: Conference Series **2793** (2024) 012008 doi:10.1088/1742-6596/2793/1/012008

is the relative velocity between the gas density and the black hole. This model can explain the relation of Rho, Cs and Mdot in the corner plot, however it can not explain the defference between the distribution of irresolvable black holes and resolvable black holes, which may help us understand the origin and formation of super massive black holes.

Table 3. Parameters description of the corner plot.

Parameters	Units	Description
Rho	-	Local gas density in the vicinity of the blackhole
Cs	-	Local gas sound-speed in the vicinity of the blackhole
Time	Second(logarithmic)	Time (scale-factor) for each entry
Mdot	$10^{10} M_{\odot}/0.978Gyr$	The mass accretion rate onto the black hole, instantaneous
Mass	$M_{\odot}(logarithmic)$	Mass of one black hole in a merger pair
Redshift	Dimensionless	-
MrdSNR	Dimensionless	SNR(LISA) of merger events in the Merger-Ringdown phase
InspSNR	Dimensionless	SNR(LISA) of merger events in the Inspiral phase
Mass ratio	-	The ratio of the mass of two black holes in a merger pair
Total mass	$M_{\odot}(logarithmic)$	The total mass of two black holes in a merger pair
Dl	Parsecs	Luminosity distance

4. Discussion and conclusions

This paper provides a new method to study the assembly of SMBHs in our universe. However, we emphasize that our model is phenomenological and that there are a number of caveats that may affect our results.

IllustrisTNG is based on the black holes seeding of $10^6 M_{\odot}$. Therefore it cannot show the different result of different seeding simulation. For example, for different BH seeds, redshift distribution, detecting ability and event rates of those GW detectors could be different, which may be able to help us understand the first SMBHs formation of the early universe.Furthermore, IllustrisTNG has limited resolution. It cannot resolve many BH parameters such as the spin parameter, which may affect the calculation of Gravitational waves. Overall, with these caveats in mind, a strong result from our calculations is that the third generation space-based GW detectors will be able to detect SMBHs in the mass range of ~ 10^6– 10^8 $M_{\odot}$ in the redshift range up to z ~ 6.

To summarize, we have calculated GW signals from SMBH mergers of a mass range ~ 10^6– 10^9 $M_{\odot}$ at redshifts z ~ 0 – 6, by post-processing the merger trees extracted from IllustrisTNG(TNG100) with a phenomenological GW model. Numerous super massive black hole mergers may occur, and these merger events may be detectable by space-based GW detectors such as LISA, Tianqin and Taiji. Space-based detectors can detect SMBHs of a wider mass range. LISA, Tianqin and Taiji will be able to detect both BBH inspirals and mergers of ~ 10^6– $10^7 M_{\odot}$ with an estimated event rate of 1 events per year at redshifts up to z ~ 6. While BBO and DECIGO may not be able to detect any super massive black hole merger event. The result of event rates in this study is in good consistency with the results in similar study[5] whose events for LISA is about two mergers of both Merger-Ringdown and Inspiral phase per year. There are some interesting parameters distribution discrepencies between resolvable SMBH mergers and irrsolvable mergers. For example, the distribution difference in local gas density may lead to the distribution difference in the mass accretion rate, which could provide some evidence on the formation of the super massive black holes. We found out that the distribution of the local gas density in the vicinity of the blackhole has two peaks, but LISA can only detect one peak from the distribution. Besides, there are also two peaks in the distribution of the mass accretion rate of total BHs, although one peak is very low and flat. Together with the distribution of Mdot and Rho, there are 3 peaks for all the supermassive black holes in the simulation, as shown in the corner plot. However, only one peak out of three can be detected, which may shed light on the mechanism of the formation of super massive black holes in the early universe.

These results demonstrate that GW detectors will be powerful tools to study SMBHs in the Universe, which will be promising to probe different BH seed models and unravel the origin of the SMBHs.

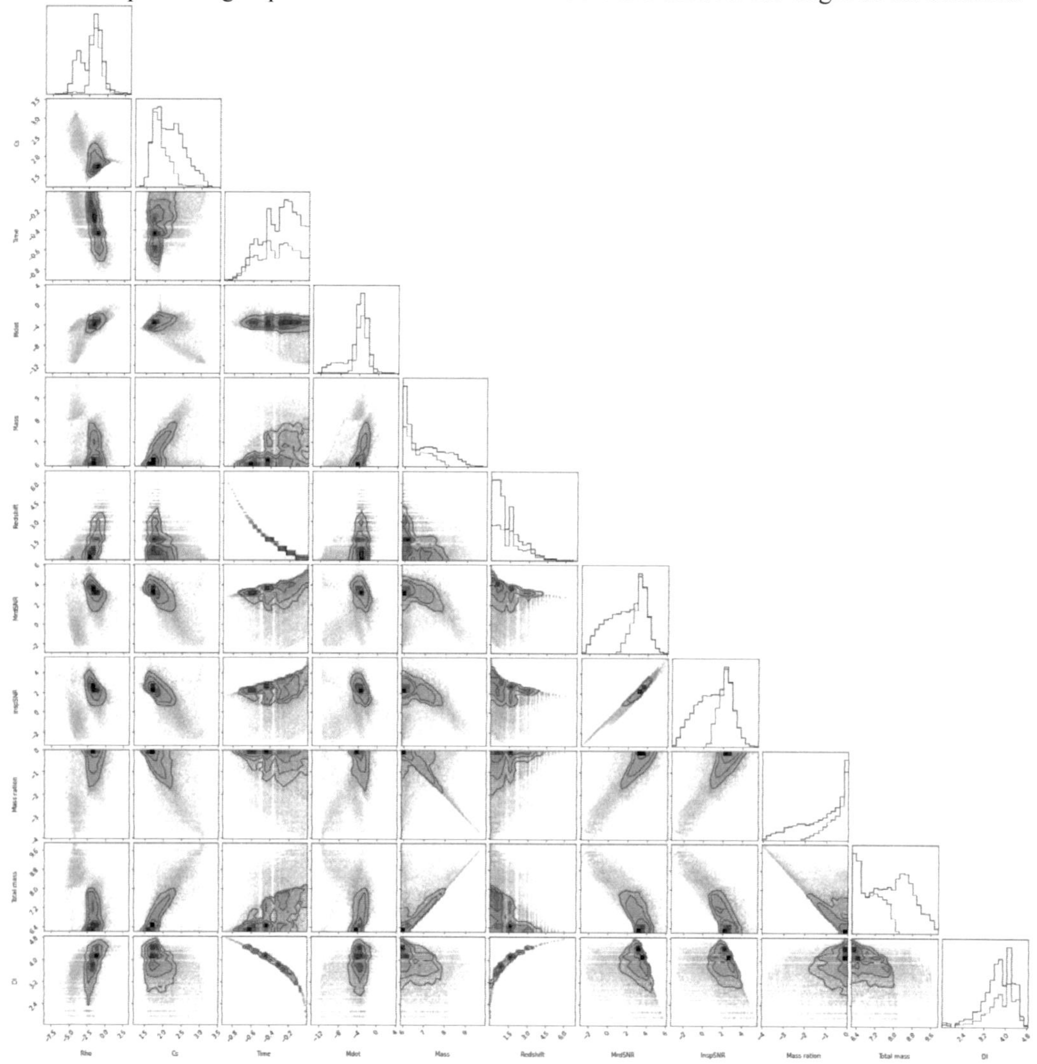

Figure 2. Corner plot of several parameters and distributions these parameters of the SMBH coalescence events from IllustrisTNG (detected by LISA). The histograms in green show the parameters distributions of irresolvable merger events while the histograms in blue show the parameters distributions of resolvable merger events. The description of the parameters can be seen in table 3.

Journal of Physics: Conference Series **2793** (2024) 012008 doi:10.1088/1742-6596/2793/1/012008

Acknowledgments
We thank Dr. Mark Hannam for providing the code of the PhenomD model for gravitational waves calculations. We thank the IllustrisTNG collaboration for sharing simulation data which constituted the analysis in the large part of this paper.

References
[1] Buonanno A and Sathyaprakash BS 2014 *Preprint* arXiv:1410.7832
[2] Amaro-Seoane P, Audley H, Babak S, Baker J, Barausse E, Bender P, Berti E, Binetruy P, Born M, Bortoluzzi D and Camp J *Preprint* arXiv:1702.00786
[3] Colpi M, Holley-Bockelmann K, Bogdanovic T, Natarajan P, Bellovary J, Sesana A, Tremmel M, Schnittman J, Comerford J, Barausse E and Berti E 2019 *Preprint* arXiv:1903.06867
[4] Haehnelt MG *Monthly Notices of the Royal Astronomical Society* 1994 **269** 199-208
[5] Salcido J, Bower RG, Theuns T, McAlpine S, Schaller M, Crain RA, Schaye J and Regan J. *Monthly Notices of the Royal Astronomical Society* 2016 **463** 870-85
[6] Nelson D, Springel V, Pillepich A, Rodriguez-Gomez V, Torrey P, Genel S, Vogelsberger M, Pakmor R, Marinacci F, Weinberger R and Kelley L *CompAC* 2019 **6** 1-29
[7] Blanchet L *Living Rev. Relativ.* 2014 17 1-87
[8] Buonanno A and Damour T *Phys. Rev. D* 1999 **59** 084006
[9] Vishveshwara CV *Nature* 1970 **227** 936-8
[10] Sathyaprakash BS and Dhurandhar SV *Phys. Rev. D* 1991 **44** 3819
[11] Finn LS and Thorne KS *Physical Review D* 2000 **62** 124021.
[12] Husa S, Khan S, Hannam M, Pürrer M, Ohme F, Forteza XJ and Bohé A *Physical Review D* 2016 **93** 044006
[13] Kumar P, Chu T, Fong H, Pfeiffer HP, Boyle M, Hemberger DA, Kidder LE, Scheel MA and Szilagyi B *Physical Review D* 2016 93 104050
[14] Barausse E, Matarrese S and Riotto A *Physical Review D* 2005 **71** 063537
[15] Moore CJ, Cole RH and Berry CP *Class Quantum Gravity* 2014 **32** 015014
[16] Mei J, Bai YZ, Bao J, Barausse E, Cai L, Canuto E, Cao B, Chen WM, Chen Y, Ding YW and Duan HZ *Prog. Theor. Phys.* 2021 **2021** 05A107
[17] Yagi K and Seto N *Physical Review D* 2011 **83** 044011
[18] Flanagan EE and Hughes SA *Physical Review D* 1998 **57** 4566
[19] Koushiappas SM and Zentner AR *Astrophys. J.* 2006 **639** 7

On propagation of gradual and impulsive Coronal Mass Ejections (CMEs)

N Mohamad Ansor[1,2], Z S Hamidi[1,2] and N N M Shariff [2,3*]

[1]Faculty of Applied Sciences, Universiti Teknologi MARA, 40450 Shah Alam, Malaysia
[2]Institute of Science, Universiti Teknologi MARA, 40450, Shah Alam, Malaysia
[3]Academy of Contemporary Islamic Studies, Universiti Teknologi MARA, 40450, Shah Alam, Malaysia

* Corresponding author's email: nnmsza@uitm.edu.my

Abstract. Coronal Mass Ejection (CME) is a significant solar activity that releases magnetic field and plasma out of the corona and they are known as the main driver of space weather. In general, CMEs are characterized as gradual and impulsive based on their kinematic properties that include velocity and acceleration. This paper is aimed to analyze the differences in behaviour of gradual and impulsive CME based on recent events (dates) in solar cycle 25 and to justify their characterizations that have been demonstrated a while ago. Data analysis is conducted by studying the height-time, height-velocity and height-acceleration profiles for each event, based on SOHO LASCO Catalogue. Results show that impulsive CME travels a lot faster than gradual CME, with average velocity of 1239 km/s and 503.6 km/s respectively. It is believed that M6.7 flare accompanying the impulsive CME has been the main reason to high velocity due to its ability to contribute more heat energy and enhances the energy conversion during magnetic reconnection. Deceleration is observed in impulsive CME that is caused by the loss of momentum as interacting with the background solar wind and small-scale acceleration is manifested by gradual CME all through the propagation.

1. Introduction

Coronal Mass Ejections (CMEs) are powerful and large explosion events in our solar system that emit a large quantity of magnetic field and energetic plasma out of the corona. Space weather is generally dominated by CMEs and many studies have proved their correlations and remarkable impacts on Earth [1], [2], [3]. As the CMEs approaching Earth's vicinity and in contact with the atmosphere, ground instruments and human high-technology activities are highly affected such as satellites breakdowns, damaged power grids and disruption of communication and navigation systems. Besides, CMEs appear to be the main cause of geomagnetic storm that is likely to be induced at high and middle latitude locations [4]. Variations in solar activity, including solar flares and CMEs are proven to have caused in changing of Earth's climate pattern considering the amount of heat produced and received by the Earth [5]. For over 25 years, CMEs have been extensively observed and studied with the aid of Large Angle and Spectrometric Coronagraph (LASCO) instrument on board the Solar and Heliospheric Observatory (SOHO) spacecraft [6]. The ejections of plasma embedded in the magnetic field, take place at an average velocity of hundreds of km/s with about 10^{16} gm of coronal plasma [7]. The plasma ejected during the events take about 1-3 days to reach the Earth and prompt consequences, depending on its speed and the proton density carried by the solar wind. Topics on different classes of CMEs have been on review by various studies that came out with different classifications and relationships between the properties [8],

[9], [10]. From an investigation on the height-speed plots of 12 CME events, they were found to be in two classifications, whereby CMEs associated with flares experienced higher velocity and low acceleration while CMEs with eruptive filament association have lower velocity and small acceleration [11]. Another study showing a parallel result was conducted years before also found that flare-associated CMEs propagated at higher speeds than CMEs with eruptive prominence association [12]. Characteristics of CMEs occurred from 1980 until 1989 were studied and the outcomes showed that 59% of CMEs that were associated with active regions exhibited constant velocity while 76% of CMEs that were associated with prominences had constant acceleration [13]. Furthermore, a study on CME events during minimum and rising phase of Solar Cycle 24 has come up with another classification – low group flare-CME and high group flare-CME [14]. Low group flare-CME are CMEs that are associated with weak flares (A, B, C) and high group flare-CME are CMEs that occur along strong flares (M and X).

As widely accepted and referred by many studies, CMEs are generally classified into two groups – gradual and impulsive. The early observations were made based on reliable instrument which is Large Angle Spectrometric Coronagraph (LASCO) onboard the Solar and Heliospheric Observatory (SOHO). A new method was formulated by [15] to model height-time plots of CMEs and incorporate them with LASCO data, and eventually came out with two distinct types of CMEs – gradual and impulsive. Gradual CMEs are recognized with their weak accelerations (less than 20 m/s^2) but lasted for a long time, up to ten of hours, going in positive acceleration. The leading edges of gradual CMEs are travelling within 400-600 km/s before leaving 30 R_{sun} and they are usually occurred along with the rising of prominences and cavities. On the other hand, impulsive CMEs are very strong accelerating events that can reach a few hundreds of m/s^2 in the low corona. Most of the times, impulsive events are correlated with solar flares and Moreton waves on visible disk as they travel at speeds beyond 750 km/s. These properties of an impulsive CME can be seen in a study on its propagation during the maximum phase of Solar Cycle 24, in April 2014. The results concluded that high energy solar flare and Moreton waves are most likely to have initiates the CME that emerged in halo structure with speed of 1600 km/s [16]. Impulsive CMEs have been on highlight due to their geoeffectiveness characteristic in causing geomagnetic storms as they mark high magnitudes in all properties. Gradual CMEs are less intense, yet they take a long time to reach the Earth which also can enhance their geoeffectiveness through long interaction with solar wind [17]. Apart from the presence of Moreton waves, impulsive CMEs are commonly linked to the solar radio bursts type II with harmonic structure and herringbone (HB) at angular widths greater than 60° and velocity greater than 500 km/s [18]. Events of CMEs are also closely related to solar radio bursts, specifically type II bursts [19], [20] as this is brought on by the eruption's magnetohydrodynamic shock. This paper is aimed to investigate the behaviour of gradual and impulsive CMEs that recently occurred in solar cycle 25 and to validate the findings on this classification which has been proposed for more than two decades back.

2. Data analysis

This study is focused on selected latest events in solar cycle 25, which began in January 2020, and it is currently in minimum phase. The following analysis is carried out by analyzing the properties of gradual and impulsive events in terms of their height-time, height-velocity and height-acceleration profiles. Several properties that are included to represent each event are height, velocity, acceleration and angular width, which all of these data are obtained from SOHO LASCO catalogue. Images of gradual CME are captured by SECCHI-A COR2 instrument, while impulsive CME are obtained from LASCO C2 observation. In presenting the properties, height-time profiles recorded and plotted by SOHO LASCO are utilized in this study. Meanwhile, velocity and acceleration data are calculated by applying the following formula [21]:

$$v = \frac{s_2 - s_1}{t_2 - t_1} \tag{1}$$

$$a = \frac{v_2 - v_1}{t_2 - t_1} \tag{2}$$

whereby s_2 is the actual position at time t_2 , s_1 is the actual position at time t_1 , t_2 is the time at point 2, t_1 is the time at point 1. Total point of data is referred to LASCO height-time data as it calculates from the first ejection of CME until it diminishes into the background solar wind. Number of velocity and acceleration data points are varied for both events as they are totally based on LASCO measurement – impulsive event has more data points than gradual event, for this study. Afterwards, all evaluated data are then plotted and curve fitted to obtain the overall trend of each properties. At the end of this study, the behaviour of propagation for both events are obtained and the differences between them are well understood.

3. Results and discussion

Gradual event is selected on 27th August 2022, an acute CME with angular width of 23° erupted at 0948 UT, lasted for almost 4 hours before easing off at 1330 UT. The CME took place at AR 3088, located at southern hemisphere of the Sun with latitude S17° at west limb of the disk. During the ejection, no surge of GOES X-ray flux recorded by NOAA at the respective active region which indicates the absent of flares associating with the CME. There were also two events of weak CME emerged before and after this gradual CME respectively and due to their insignificant properties, there were no geomagnetic storms triggered from the events. On the other hand, an event on 28th August 2022 is selected to represent impulsive CME that began at 1612 UT, lasted for more than 3 hours and ended at 1942 UT. For this event, an M6.7 flare was associated as it emitted within the timeframe of the CME at 1548 UT, reached its peak at 1619 UT and settled down at 1646 UT. Due to high class flare association, this CME is seemed to be stronger to appear in a full halo structure compared to gradual CME with no flares and led to smaller angular width. Figure 1 (first row) show evolution of gradual CME on 27th August 2022 from the view of SECCHI-A COR2 and evolution of impulsive CME on 28th August 2022 as seen in LASCO C2 field of view (second row).

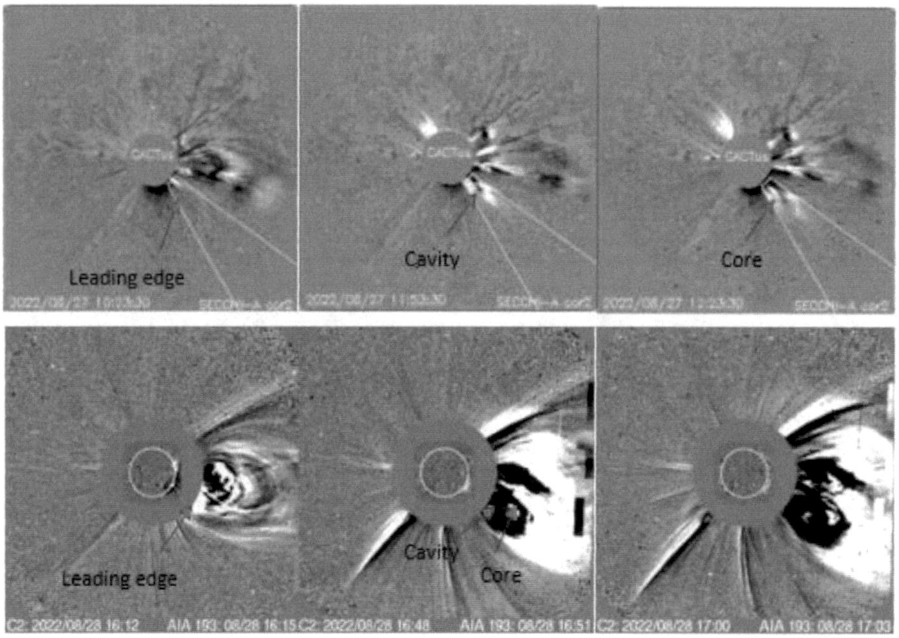

Figure 1. Images in first row demonstrate the evolution of gradual CME on 27th August 2022 observed by SECCHI-A COR2. Displayed in the second row are the evolution of impulsive CME on 28th August 2022 from SOHO LASCO C2 field of view.

Journal of Physics: Conference Series **2793** (2024) 012009 doi:10.1088/1742-6596/2793/1/012009

As stated in the database, gradual CME started its initiation at 0948 UT, yet its leading edge is apparently revealed in SECCHI-A COR2 view at 1023 UT, 35 minutes later. Leading edge in most cases is seen to expand and brighten, before lifting off the Sun's surface while forming a loop-like structure, which eventually reconnects with the Sun. After more than an hour, cavity of the CME is seen following the leading edge and the core is successfully released at 1223 UT, precisely 2 hours after the emergence of leading edge. Subsequently, the CME proceeds its propagation further from the Sun for another an hour before emerging with background solar wind at 1330 UT. On the other situation, SOHO LASCO C2 recorded the beginning of impulsive CME to be at 1612 UT, by which its leading edge is also coming into sight at that moment, pictured in the first image of second row. As this event is characterized as impulsive event, the whole structure of CME is manifestly moving faster than gradual and that leads to surfacing of cavity and its core in 36 minutes after the initiation. Throughout its propagation, impulsive event took more than 3 hours to combine with ambient solar wind, almost similar with gradual CME duration with a few minutes difference. By observing the brightness of both CME structures, impulsive event exhibits more prominent and brighter leading edge and core, while its cavity is wider in radius and has a darker void compared to gradual structure. From the concepts, it is known that cavity correlates with low density region, located close to coronal base in the quiet streamer and the bright featured core is believed to be dense materials originated from prominences in the streamer cavity before CME is ejected [22]. Hence, the differences in the features of three-part structure in both events are assumed to demonstrate the condition of the density at that particular part – impulsive CME has larger mass of dense materials in the core during the ejection and the increase in magnetic force between electric current in the core and opposite eddy current which projects the plasma away from the core, leaving wider radius of cavity.

Solar flare association is only observed during impulsive event, that the M6.7 flare began a few minutes before CME emergence at 1548 UT, reached its peak flux at 1619 UT and ended at 1646 UT. Meanwhile during gradual CME, no presence of flares is observed which is presumed to be resulting the slow propagation and unspectacular features. Figure 2 shows the properties of gradual and impulsive CME consist of height-time, height-velocity and height-acceleration profiles. Based on height-time data, both events started to lift off the Sun's surface at nearly similar height at 2.77 R_{sun} and 2.84 R_{sun}, for gradual and impulsive respectively. As the CMEs are manoeuvring into interplanetary space, the height change rate for impulsive CME records smaller value, indicating it is moving more gradually with its graph demonstrates less steep than gradual height-time graph. Considering both events lasted for about the same duration, impulsive CME covers more distance and travels further, reaching 26.08 R_{sun} while gradual CME marks its maximum height achieved at 11.88 R_{sun}, less than half from the impulsive peak height. Velocity trends portrayed by both events are definitely opposite to each other, whereby gradual CME experiences slow increment in velocity towards positive direction with a slight of acceleration observed at a point and impulsive CME undergoes linear deceleration as its velocity constantly decreases throughout the propagation. According to the measurement method, average velocity of gradual CME is obtained at 503.6 km/s, achieves its highest velocity at 526 km/s and is assumed to proceed with constant velocity beyond 12 R_{sun}. The range of velocity is more confined than impulsive CME, going from minimum velocity at 470 km/s and up to 526 km/s, which takes 3 hours 42 minutes to cover 9.11 R_{sun} of height and 56 km/s of speed difference. Dissimilar to impulsive CME, its velocity range is distinctly broader as it goes from 1375 km/s and keeps falling towards negative direction before reaching the end point with 1120 km/s. Over the 3 hours and 30 minutes of its propagation, the impulsive CME accomplishes height increment of 23.24 R_{sun} from starting to ending with 255 km/s of velocity difference, that is considerably fast-moving.

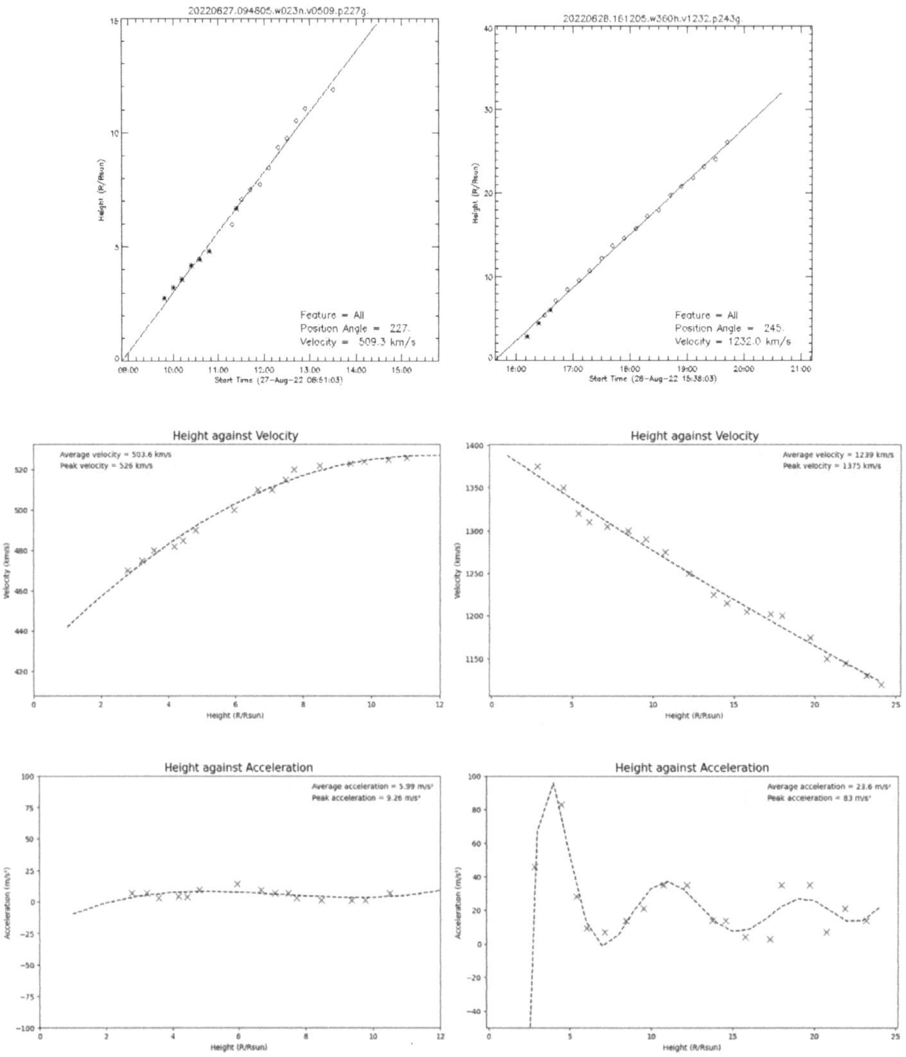

Figure 2. Height-time profile (obtained from SOHO LASCO Catalogue), height-velocity profile and height-acceleration profile for gradual CME (left column) and impulsive CME (right column).

Overall acceleration trend is more prominent with a few fluctuations seen during the propagation of impulsive CME, while gradual CME manifests a very low acceleration which all points are marked close to 0 m/s^2. At initial state of gradual CME, magnetic field with embedded plasma accelerates at 6.94 m/s^2 and fluctuate before hitting its highest acceleration at 1118 UT with 14 m/s^2 and back in oscillating towards deceleration manner. The evaluated average acceleration marks at a very low level with 5.99 m/s^2 and towards positive direction. Prominent differences can be observed in height-acceleration profile of impulsive CME, demonstrating unsteady trend with several up and down phases that sustain until the end. Average acceleration is obtained at 23.6 m/s^2, generally going towards negative manner with peak acceleration of 83 m/s^2, relatively higher than gradual CME. At the beginning of impulsive CME, magnetic field and plasma are accelerated at a very high magnitude as demonstrated by the peak graph before strikingly decelerates and continues oscillating in a persistent behaviour. Within the timeframe of these two events, flare of M6.7 occurred concurrently with impulsive CME and gradual CME emerged on its own without associated flares. The existence of associated flare can be seen its effects on the CMEs properties, specifically in terms of their velocity and acceleration, which evidently contradict.

On the whole, gradual CME steadily increases in velocity until it achieves constant speed beyond 12 R_{sun} and impulsive CME decreases its velocity linearly from the beginning until the end. The change in acceleration of gradual CME is not really significant with minor velocity increments observed, while impulsive CME exhibits a very unstable trend with peak acceleration recorded at the beginning phase of ejection and decelerates afterwards. Formerly, it was well understood that solar flares could also compliment to CME energy which results to the powerful condition upon magnetic field and plasma ejection. By way of magnetic reconnection process, the flare energy is induced by a sudden discharge of free magnetic energy. In the time of discharge, the conversion process of magnetic field energy into heating plasma has been proven to be the cause of CME velocity variation [23]. As flares have the ability to magnify heat energy to CMEs, the magnetic field that is being ejected also has its energy boosted by the associated flares. Therefore, flares with high energy (specifically M- and X- class flares) would distribute more heat energy to CMEs and increases the conversion process of magnetic field into heating plasma and accelerating particles, leads to CMEs that move at high speeds.

Height-acceleration profiles are as well in a disparate trend – small hints of acceleration for gradual CME and fluctuating towards negative direction for impulsive CME. Many studies have seen the same behaviour on both types of CME whereby fast-moving CMEs are prone to decelerate and slow-moving CMEs are found to be accelerating upon contacting with background solar wind [15], [24]. In general, all CMEs with or without the presence of flares are driven by main factors which are Lorentz force, magnetic tension, Sun's gravity and solar wind drag force. The similar ending phase that is experienced by all CMEs are the incorporation with ambient solar wind and move together at solar wind speed. On the ejection of high-speed CMEs, they contain a huge momentum while propagating, which later is being transferred to solar wind and results to deceleration of CMEs as a loss of momentum. When this transfer is lasted for a long period, CMEs eventually lose a big portion of their momentum and are brought to decreasing in speed as approaching the end phase. These findings are correlated with several studies which also found the acceleration and deceleration behaviour on different type of CMEs [17], [25]. Therefore, the deceleration experienced by impulsive CME on 28th August 2022 is believed to be caused by the loss in momentum into the solar wind during their propagation. Meanwhile, gradual CME is not showing greater a change in velocity due to the absent of associated flare that could provide more heat energy and increase the conversion process during ejection.

4. Conclusion

In summary, the properties of gradual and impulsive CME are presented based on their height-time, height-velocity and height-acceleration profiles. By analyzing the behaviour of each event, it can be inferred that impulsive CME is travelling more rapidly than gradual CME with its average velocity is relatively higher. However, it encounters deceleration throughout its propagation due to the loss of momentum which has been transferred to background solar wind. As impulsive CME is ejected at a very high speed and is generally fast, it covers more distance per minute with rate of change of height is also higher. The presence of associated flares is believed to have caused the difference in behaviour for both events. Impulsive CME with higher average velocity is technically enhanced by the associated M6.7 flare, in a way that the strong flare gives out more heat energy to the CME during energy conversion process which later leads to soaring initial velocity and wider angular width. In opposition, gradual CME without associated flare is emerged in low average velocity with minimal acceleration and small angular width.

Acknowledgments
We would like to thank STEREO, LASCO, SDO/AIA, NOAA, NASA and SWPC for providing us the free-access online databases. This study was partially supported by UiTM research grant, 600-RMC/YTR/5/3 (013/2020) and FRGS/1/2023/STG07/UITM/02/11 under Kementerian Pengajian Tinggi Malaysia.

References

[1] R. Pirjola, "Space weather effects on power grids," in *Space Weather- Physics and Effects*, Berlin, Heidelberg: Springer Berlin Heidelberg, 2007, pp. 269–288. doi: 10.1007/978-3-540-34578-7_10.

[2] P. L. Verma, P. Singh, and P. Singh, "Coronal Mass Ejections And Disturbances In Solar Wind Plasma Parameters In Relation With Geomagnetic Storms," *J Phys Conf Ser*, vol. 511, p. 012060, May 2014, doi: 10.1088/1742-6596/511/1/012060.

[3] N. Gopalswamy, "Halo coronal mass ejections and geomagnetic storms," *Earth, Planets and Space*, vol. 61, no. 5, pp. 595–597, May 2009, doi: 10.1186/BF03352930.

[4] N. Mohamad Ansor and Z. S. Hamidi, "Effects of CME-induced Geomagnetic Storm on Geomagnetic Induced Current at High and Middle Latitudes," in *Journal of Physics: Conference Series*, Institute of Physics, 2022. doi: 10.1088/1742-6596/2287/1/012035.

[5] N. Mohamad Ansor, Z. S. Hamidi, and N. N. M. Shariff, "The Impact on Climate Change Due to the Effect of Global Electromagnetic Waves of Solar Flare and Coronal Mass Ejections (CMEs) Phenomena," in *Journal of Physics: Conference Series*, Institute of Physics Publishing, Sep. 2019. doi: 10.1088/1742-6596/1298/1/012019.

[6] G. E. Brueckner *et al.*, "The Large Angle Spectroscopic Coronagraph (LASCO)," *Sol Phys*, vol. 162, no. 1–2, pp. 357–402, Dec. 1995, doi: 10.1007/BF00733434.

[7] D. F. Webb and T. A. Howard, "Coronal Mass Ejections: Observations," *Living Rev Sol Phys*, vol. 9, 2012, doi: 10.12942/lrsp-2012-3.

[8] G. S. Suryanarayana, "Correlations between the CME acceleration, other CME parameters and flare energy," *J Atmos Sol Terr Phys*, vol. 185, pp. 1–6, Apr. 2019, doi: 10.1016/j.jastp.2019.01.014.

[9] R. Mawad, W. Abdel-Sattar, and H. M. Farid, "An association of CMEs with solar flares detected by Fermi γ-ray burst monitor during solar cycle 24," *New Astronomy*, vol. 82, p. 101450, Jan. 2021, doi: 10.1016/j.newast.2020.101450.

[10] J. Nicewicz and G. Michalek, "Classification of CMEs Based on Their Dynamics," *Sol Phys*, vol. 291, no. 5, pp. 1417–1432, May 2016, doi: 10.1007/s11207-016-0903-4.

[11] R. M. MacQueen and R. R. Fisher, "The kinematics of solar inner coronal transients," *Sol Phys*, vol. 89, no. 1, pp. 89–102, Nov. 1983, doi: 10.1007/BF00211955.

[12] J. T. Gosling, E. Hildner, R. M. MacQueen, R. H. Munro, A. I. Poland, and C. L. Ross, "The speeds of coronal mass ejection events," *Sol Phys*, vol. 48, no. 2, pp. 389–397, Jun. 1976, doi: 10.1007/BF00152004.

[13] O. C. St. Cyr, J. T. Burkepile, A. J. Hundhausen, and A. R. Lecinski, "A comparison of ground-based and spacecraft observations of coronal mass ejections from 1980–1989," *J Geophys Res Space Phys*, vol. 104, no. A6, pp. 12493–12506, Jun. 1999, doi: 10.1029/1999JA900045.

[14] N. M. Ansor, Z. S. Hamidi, and N. N. M. Shariff, "Characteristics of Different Groups of Flare-CME in the Minimum to Rising Phase of Solar Cycle 24," *Sains Malays*, vol. 52, no. 3, pp. 981–992, Mar. 2023, doi: 10.17576/jsm-2023-5203-21.

[15] N. R. Sheeley, J. H. Walters, Y. -M. Wang, and R. A. Howard, "Continuous tracking of coronal outflows: Two kinds of coronal mass ejections," *J Geophys Res Space Phys*, vol. 104, no. A11, pp. 24739–24767, Nov. 1999, doi: 10.1029/1999JA900308.

[16] Z. S. Hamidi and N. N. M. Shariff, "The Propagation of an Impulsive Coronal Mass Ejections (CMEs) due to the High Solar Flares and Moreton Waves," *International Letters of Chemistry, Physics and Astronomy*, vol. 33, pp. 118–126, May 2014, doi: 10.18052/www.scipress.com/ILCPA.33.118.

[17] M. Temmer *et al.*, "CME propagation through the heliosphere: Status and future of observations and model development," *Advances in Space Research*, Jul. 2023, doi: 10.1016/j.asr.2023.07.003.

[18] Z. S. Hamidi, N. H. Zainol, N. N. M. Shariff, and N. M. Ansor, "Solar Radio Burst Type II Characteristics and Coronal Mass Ejections (CMEs) Structure Based on the Presence of a

Moreton Wave," *Sains Malays*, vol. 51, no. 4, pp. 1261–1269, Apr. 2022, doi: 10.17576/jsm-2022-5104-24.

[19] S. N. A. Shamsuddin, N. M. Ansor, Z. S. Hamidi, and N. N. M. Shariff, "An analysis of slow drift type II radio burst and its association with full halo coronal mass ejection on 12th January 2022," in *Journal of Physics: Conference Series*, Institute of Physics, 2023. doi: 10.1088/1742-6596/2543/1/012013.

[20] N. M. Ansor, Z. S. Hamidi, and N. N. M. Shariff, "An automated system in detecting solar radio bursts type II and IV associated to multiple Coronal Mass Ejections," in *Journal of Physics: Conference Series*, Institute of Physics Publishing, Dec. 2019. doi: 10.1088/1742-6596/1411/1/012015.

[21] Solar and Heliospheric Observatory, "Measuring the Motion of a Coronal Mass Ejection."

[22] N. Mittal and U. Narain, "Initiation of CMEs: A review," *J Atmos Sol Terr Phys*, vol. 72, no. 9–10, pp. 643–652, Jun. 2010, doi: 10.1016/j.jastp.2010.03.011.

[23] R. Jain, M. Aggarwal, and P. Kulkarni, "Relationship between CME dynamics and solar flare plasma," *Res Astron Astrophys*, vol. 10, no. 5, pp. 473–483, May 2010, doi: 10.1088/1674-4527/10/5/007.

[24] N. Gopalswamy, A. Lara, R. P. Lepping, M. L. Kaiser, D. Berdichevsky, and O. C. St. Cyr, "Interplanetary acceleration of coronal mass ejections," *Geophys Res Lett*, vol. 27, no. 2, pp. 145–148, Jan. 2000, doi: 10.1029/1999GL003639.

[25] F. Shen, S. T. Wu, X. Feng, and C.-C. Wu, "Acceleration and deceleration of coronal mass ejections during propagation and interaction," *J Geophys Res Space Phys*, vol. 117, no. A11, p. n/a-n/a, Nov. 2012, doi: 10.1029/2012JA017776.

Journal of Physics: Conference Series **2793** (2024) 012010 doi:10.1088/1742-6596/2793/1/012010

Multi-objective optimal design for the railway bogie frame system under stress and fatigue analysis

Cheng-Kang Lee[1], Chien-Kuan Liu[2] and Yung-Chang Cheng[3,*]

[1]Department of Industrial Engineering and Management, Cheng Shiu University, Kaohsiung, Taiwan.

[2]Department of Mechatronics Engineering, National Kaohsiung University of Science and Technology, Kaohsiung, Taiwan.

[3]Department of Mechatronics Engineering, National Kaohsiung University of Science and Technology, Kaohsiung, Taiwan.

* Corresponding author's email: yccheng@nkust.edu.tw

Abstract. This work focuses on applying a novel and integrated multi-objective optimization approach to enhance the strength of the railway bogie frame system under both static loads and fatigue tests. The finite element modelling is conducted using ANSYS/Workbench software, and the von Mises stress and fatigue safety factor of the bogie frame are computed through conventional simulations for EN 13749 and EN 15663 testing. After employing the uniform design method, the fatigue safety factor is increased from 1.924 to 2.586, resulting in a 34.41% improvement rate for the bogie frame. Additionally, the von Mises stress is reduced from 56.96 MPa to 42.91 MPa, reflecting a 24.66% improvement rate compared to the original design. Following the implementation of the multi-objective optimum design process, the fatigue safety factor further improves from 1.924 to 2.742, resulting in a 42.51% enhancement for the bogie frame. Simultaneously, the von Mises stress decreases from 56.96 MPa to 40.48 MPa, yielding an improvement rate of 28.92%.

1. Introduction

With the advancement of science and technology and a substantial increase in computer processing power, numerous academics are now turning to simulation analysis to address engineering challenges associated with railway vehicles. The strength design and service life of bogies stand as critical factors in guaranteeing the secure operation of railway cars on tracks.

A railway vehicle's bogie underwent stress and fatigue studies in multiple investigations. Liliana et al. [1] examined the distribution of stresses and strains in a bogie frame. Kuczyk et al. [2] conducted strength analysis on specific bogie components, such as the frame and wheelset axle, following the PN-EN 13749 standard. Abdurasulov et al. [3] theoretically analyzed the stress-strain condition of the industrial traction unit PE2U and PE2M frame using finite element analysis (FEA). Seo et al. [4] investigated the bogie frame's fatigue strength to explore the impact of a multiaxial load. An et al. [5] analyzed the metro bogie frame's fatigue strength under random load. Lutema and Mohammedseid [6] estimated the bogie's fatigue life through uncertainty-parameter-based cumulative damage calculations.

Several studies have addressed the static stress analysis of a railway bogie frame system based on the literature. However, there has been a lack of literature demonstrating optimization design for a model with a bogie frame. Consequently, this article introduces a series of simulation experiments employing the uniform design of experiments. According to standard simulations for EN 13749 and EN 15663, the

Journal of Physics: Conference Series **2793** (2024) 012010 doi:10.1088/1742-6596/2793/1/012010

maximum von Mises stress and minimum fatigue safety factor for a bogie frame model are determined by ANSYS/Workbench software. Ultimately, by integrating the genetic algorithm (GA), entropy weighting analysis (EWA), grey relation analysis (GRA), and Kriging interpolation (KGI) approach, the multi-objective optimum design procedure has generated an improved design version for a bogie frame.

2. Finite element analysis for railway bogie system

2.1. 3D model of railway bogie

The 3D model of a railway bogie frame is reconstructed by SolidWorks software, as depicted in Figure 1. In this 3D model, certain geometric shapes unrelated to structural strength are omitted from the original design.

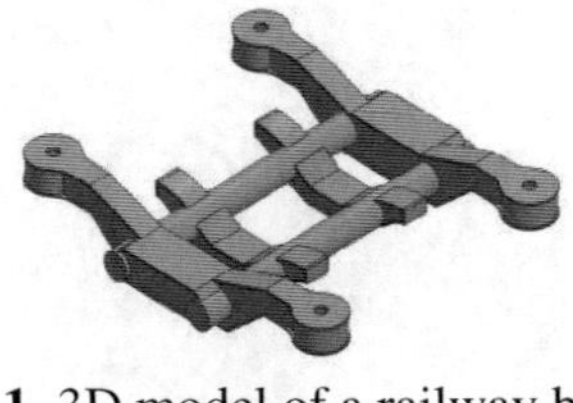

Figure 1. 3D model of a railway bogie frame.

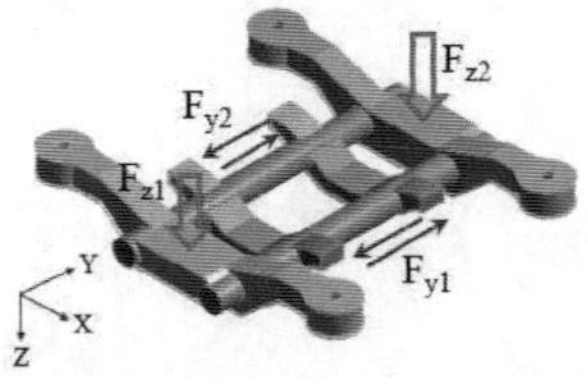

Figure 2. External forces on bogie frame.

2.2. Static stress analysis

In the static stress analysis for the bogie frame simulation testing, external forces F_z and F_y are applied to the frame in accordance with the EN 13749 [7] and EN 15663 [8] standards, as illustrated in Figure 2. As per the EN 13749 and EN 15663 test standards, the vertical and lateral forces, F_z and F_y, are determined by the following equations:

$$F_{z1\max} = F_{z2\max} = {F_{z\max}}/{2} = {1.4g\left(M_v + P_1 + m^+\right)}/{4} \tag{1a}$$

$$F_{y1\max} = F_{y2\max} = {F_{y\max}}/{2} = 10^4 + {(M_v + P_1)g}/{12}, \tag{1b}$$

where M_v denotes the weight of a car body and P_1 denotes the weight of full passengers. m^+ denotes the bogie frame weight. According to EN 13749 testing standard, there are 9 combinations of load modes as shown in Table 1(a). In this study, it is the bogie of a passenger car, $\alpha = 0.1$ and $\beta = 0.2$ are set and utilized in Table 1(a). Following the equations (1a) and (1b), the nine external force combinations actually applied to the bogie can be obtained, as shown in Table 1(b).

Table 1. The external forces for EN 13749 standard.

(a)

Load case	F_{z1}	F_{z2}	F_y
1	$F_z/2$	$F_z/2$	0
2	$(1+\alpha-\beta)\,F_z/2$	$(1-\alpha-\beta)\,F_z/2$	0
3	$(1+\alpha-\beta)\,F_z/2$	$(1-\alpha-\beta)\,F_z/2$	$+F_y$
4	$(1+\alpha+\beta)\,F_z/2$	$(1-\alpha+\beta)\,F_z/2$	0
5	$(1+\alpha+\beta)\,F_z/2$	$(1-\alpha+\beta)\,F_z/2$	$+F_y$
6	$(1-\alpha-\beta)\,F_z/2$	$(1+\alpha-\beta)\,F_z/2$	0
7	$(1-\alpha-\beta)\,F_z/2$	$(1+\alpha-\beta)\,F_z/2$	$-F_y$
8	$(1-\alpha+\beta)\,F_z/2$	$(1+\alpha+\beta)\,F_z/2$	0
9	$(1-\alpha+\beta)\,F_z/2$	$(1+\alpha+\beta)\,F_z/2$	$-F_y$

(b)

Load case	F_{z1}	F_{z2}	F_y
1	64000	64000	0
2	57600	44800	0
3	57600	44800	50572
4	83200	70400	0
5	83200	70400	50572
6	44800	57600	0
7	44800	57600	-50572
8	70400	83200	0
9	70400	83200	-50572

As depicted in Figure 3, the boundary condition for the bogie frame is presented. The bogie frame is constructed from SMA490BW material, and its mechanical properties are detailed in Table 2 [9]. Mesh convergence results obtained using ANSYS/Workbench, an FEA program, are shown in Figure 4(a). The optimal element size, determined from the mesh convergence findings, is 35, and the load case corresponds to the fourth case. Additionally, Figure 4(b) illustrates the distribution of the von Mises stress in the bogie frame model, with a maximum von Mises stress of 56.96 MPa.

Table 2. Mechanical property of a bogie frame.

Body	Material	Young's Modulus (GPa)	Poisson's Ratio	Density (kg/m^3)
Bogie Frame	SMA490BW	200	0.323	7700

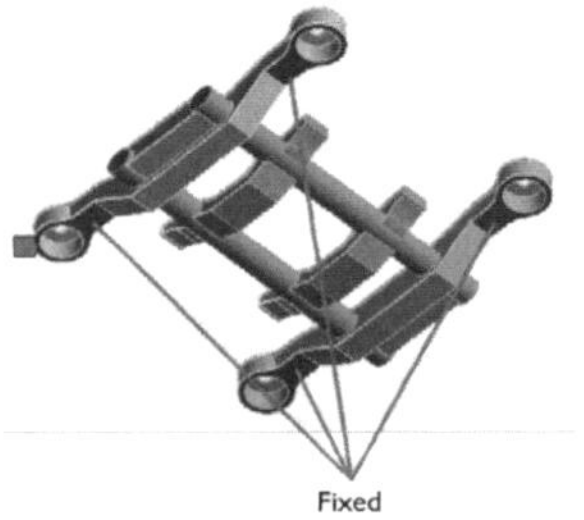

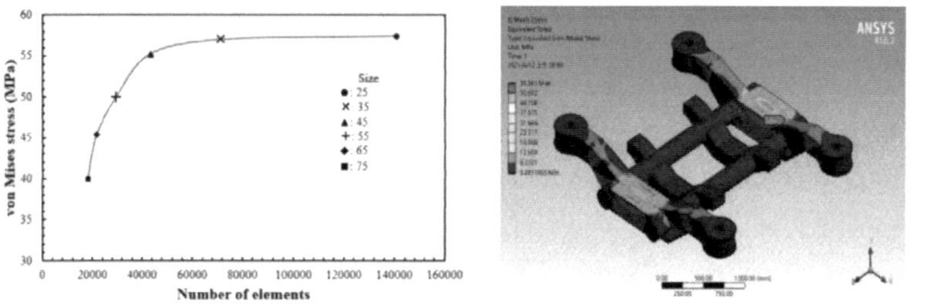

(a) (b)

Figure 3. Boundary condition. **Figure 4**. (a) Mesh convergence and (b) von Mises stress results.

2.3. Fatigue safety analysis

The EN 13749 test standard is used in this study to demonstrate the fatigue safety analysis of a railway bogie frame model. The vertical and lateral forces applied in the first stage of the fatigue test are: $F_s + F_{qs} + F_d$ respectively. The vertical and lateral forces applied in the second stage of the fatigue test are: $F_s + 1.2\left(F_{qs} + F_d\right)$. In the fatigue test The vertical and lateral forces applied in the third stage of the test are: $F_s + 1.4\left(F_{qs} + F_d\right)$ respectively, as shown in Figure 5. As seen in Figure 6, apply external force to the bogie's transverse and vertical axes. The following formulas are used to compute the force in the lateral (y) and vertical (z) directions in accordance with the EN 13749 fatigue test standard:

Static: $F_{z1s} = F_{z2s} = F_z/2$ (2a)

Quasi-static: $F_{z1qs} = F_{z2qs} = \pm\alpha F_z/2$ (2b)

Dynamic: $F_{z1d} = F_{z2d} = \pm\beta F_z/2$ (2c)

Quasi-static: $F_{y1qs} = F_{y2qs} = \pm 0.063\left(F_z + m^+ g\right)$ (3a)

Dynamic: $F_{y1d} = F_{y2d} = \pm 0.063\left(F_z + m^+ g\right)$ (3b)

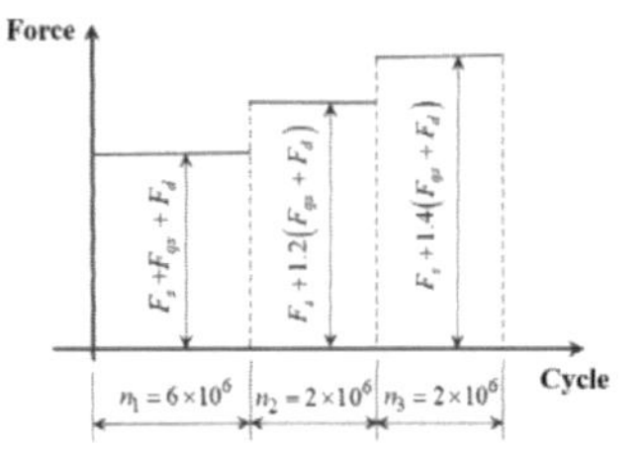

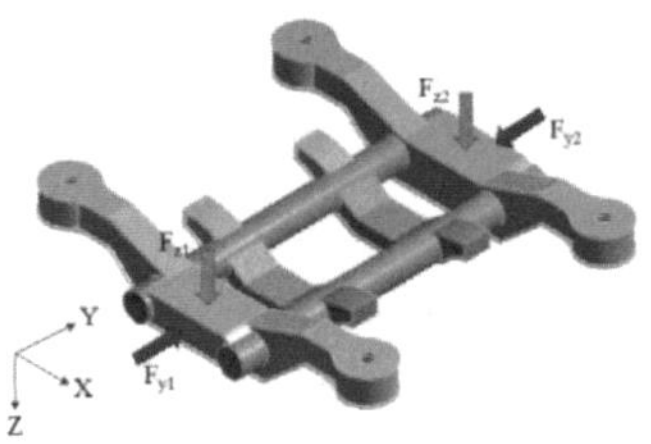

Figure 5. The loads history in fatigue testing. **Figure 6**. External loads in fatigue testing.

As depicted in Figure 7, the SN curve is required for evaluating the fatigue safety factor in ANSYS/Workbench. Figure 8 depicts the determination of mesh convergence and fatigue safety factor, considering the boundary condition shown in Figure 3. The fatigue safety factor is 1.924, and the optimal element size is determined to be 35.

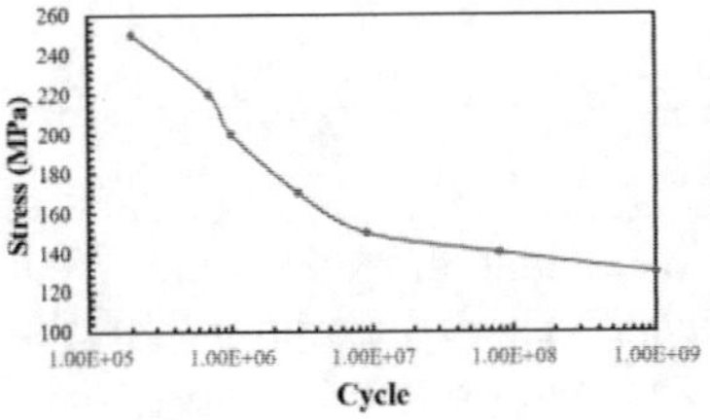

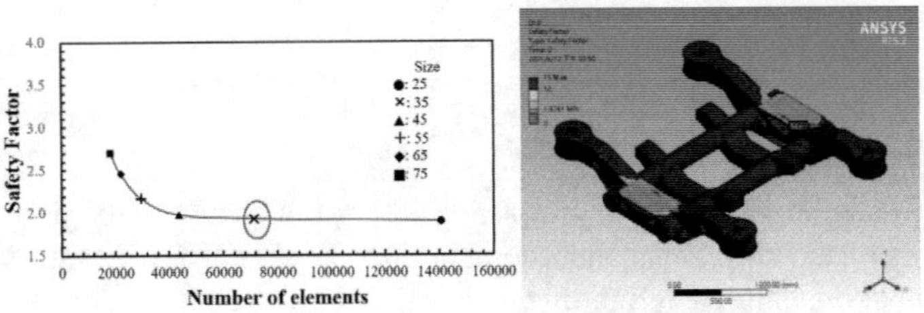

(a) (b)

Figure 7. SN curve of a bogie frame. **Figure 8**. (a) Mesh convergence and (b) von Mises stress results.

3. Optimal design analysis

3.1. Uniform design of experiment

Five control elements are chosen for the railway bogie frame model, as illustrated in Figure 9: thickness of the frame (AT), width of both sides of the frame (2-L), width of the center of the frame (3-L), height of the front side of the frame (4-H), and length of both sides of the frame (5-L). In Table 3, the basic values, lower and upper bounds for each control factor are presented. The design space is continuous as all elements are continuous. To generate the experiment simulations, the uniform design technique proposed by Fang and Wang [10] can be applied. Table 4(a) shows the uniform table U16(16^{12}) with 16 rows [8]. Table 4(b) displays each experiment simulation of a railway bogie frame system. 3D models for each design are created using SolidWorks, and the minimum fatigue safety factor and maximum von Mises stress are computed using ANSYS/Workbench. The 14th simulation exhibits the highest improvement rate when compared to the original design in Table 4(b). Consequently, the 14th model may be considered the improved version resulting from the uniform design strategy.

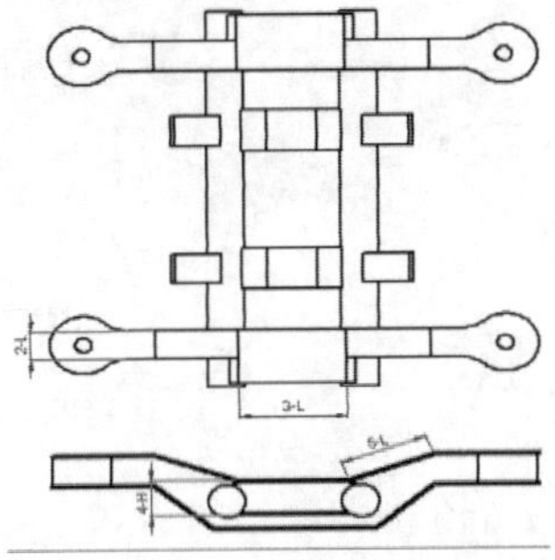

Figure 9. Control factors for bogie frame model

Table 3. Design ranges for control factors

Notation	Lower bound (mm)	Basic value (mm)	Upper bound (mm)
AT	13.50	15.00	16.50
2-L	194.75	205.00	215.25
3-L	617.50	650.00	682.50
4-H	209.00	220.00	231.00
5-L	503.25	529.74	556.23

Table 4. Uniform table U16(16^{12})

(a)

No.	AT (mm)	2-L (mm)	3-L (mm)	4-H (mm)	5-L (mm)
0	15	205	650	220	529.74
1	13.50	200.22	647.83	219.27	545.63
2	13.70	207.05	682.50	231.00	531.51
3	13.90	213.88	643.50	217.80	517.38
4	14.10	197.48	678.17	229.53	503.25
5	14.30	204.32	639.17	216.33	549.16
6	14.50	211.15	673.83	228.07	535.04
7	14.70	194.75	634.83	214.87	520.91
8	14.90	201.58	669.50	226.60	506.78
9	15.10	208.42	630.50	213.40	552.70
10	15.30	215.25	665.17	225.13	538.57
11	15.50	198.85	626.17	211.93	524.44
12	15.70	205.68	660.83	223.67	510.32

(b)

No.	Min. fatigue Safety Factor	Improvement	Max. von-Mises Stress (MPa)	Improvement
0	1.488	-22.64%	73.068	-28.28%
1	1.491	-22.49%	71.368	-25.29%
2	1.451	-24.59%	74.582	-30.94%
3	1.787	-7.13%	60.985	-7.06%
4	1.703	-11.49%	64.194	-12.70%
5	1.681	-12.61%	64.374	-13.01%
6	1.968	2.28%	55.432	2.68%
7	2.064	7.29%	53.110	6.76%
8	1.875	-2.53%	58.460	-2.63%
9	1.889	-1.79%	57.669	-1.24%
10	2.342	21.70%	47.196	17.14%
11	2.211	14.90%	49.693	12.76%
12	2.113	9.80%	51.876	8.93%

13	15.90	212.52	621.83	210.47	556.23
14	16.10	196.12	656.50	222.20	542.10
15	16.30	202.95	617.50	209.00	527.97
16	16.50	209.78	652.17	220.73	513.85

13	2.586	34.41%	42.913	24.66%
14	2.471	28.41%	44.960	21.07%
15	2.451	27.37%	45.017	20.97%
16	1.488	-22.64%	73.068	-28.28%

3.2. Multi-objective optimal design analysis

To achieve the multi-goal optimization design for a railway bogie frame model, this study employs an innovative technique that incorporates the use of uniform design tables, genetic algorithms, Kriging interpolation, entropy weighting analysis, and grey relation analysis. The optimization design approach employed to obtain the optimal design is illustrated in Figure 10 [11]. Upon completing this optimization process, Table 5 presents the ultimate ideal outcomes. Additionally, Figure 11 depicts the minimum fatigue safety factor of 2.742 and the maximum von Mises stress of 40.48 MPa for the ideal railway bogie frame model design. Finally, Table 6 illustrates the improvement in two objectives.

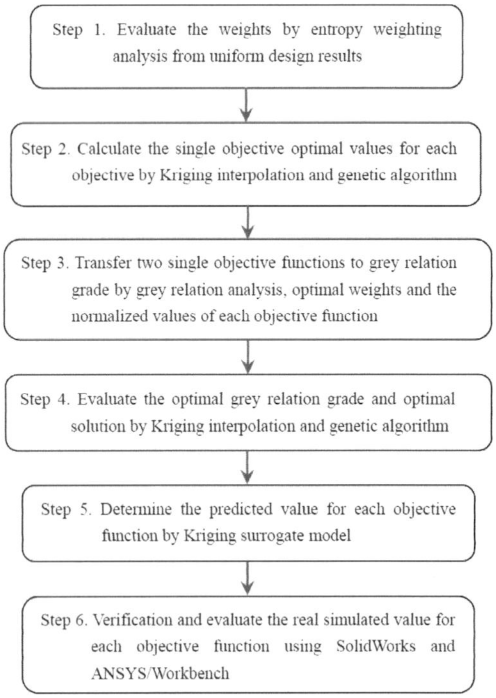

Figure 10. Multiple objective optimal design process.

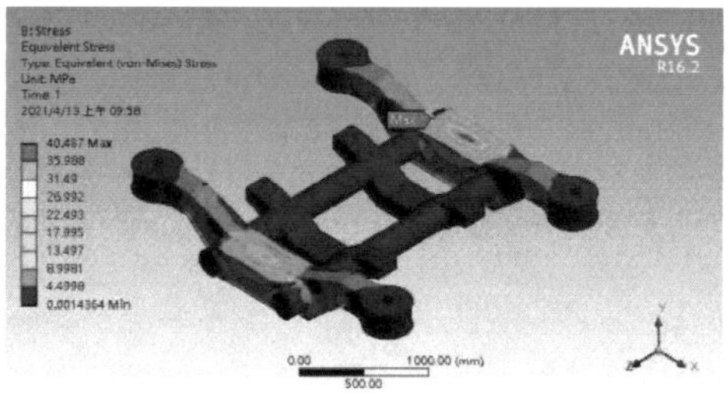

(a)

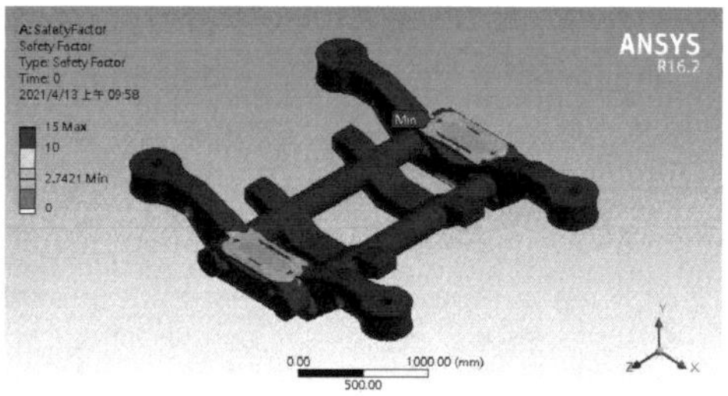

(b)

Figure 11. The (a) maximum von Mises stress and (b) the fatigue safety factor for the final optimal design for the EN 13749 static load and fatigue simulations.

Table 5. Optimal solution and value.

Optimal Solution					Optimal Value	
AT(mm)	2-L(mm)	3-L(mm)	4-H(mm)	5-L(mm)	Safety Factor	von Mises Stress(MPa)
16.5	194.75	682.5	231	503.25	2.742	40.487

Table 6. FEA values and improvement rate for various phases.

Phase	Measure	Value	Improvement (%)
Original design	Fatigue Safety Factor	1.924	—
	Maximum von Mises Stress(MPa)	56.96	—
After uniform experiments	Fatigue Safety Factor	2.586	34.41%
	Maximum von Mises Stress(MPa)	42.91	24.66%
After Multi-objective optimization	Fatigue Safety Factor	2.742	42.51%
	Maximum von Mises Stress(MPa)	40.48	28.92%

Journal of Physics: Conference Series **2793** (2024) 012010 doi:10.1088/1742-6596/2793/1/012010

4. Conclusion

This study has incorporated UD, KGI, EWA, GRA, and GA to carry out the improvement design for a bogie frame model under static loading and fatigue testing simulations. In comparison to the original design, the multi-objective optimum design technique resulted in a maximum improvement of 42.51% and 28.92% for the von Mises stress and fatigue safety factor, respectively. With its enhanced design, the railway bogie frame can now successfully meet the EN13749 test standard, thereby increasing its strength and safety.

Acknowledgments

The authors would like to express our gratitude and appreciation for the financial support of the Ministry of Science and Technology (MOST) under grant no's. 111-2221-E-992-057.

References

[1] Liliana R C A, Florin B, Nicolae I and Costica A 2015 Stresses in a bogie frame of a rail carriage. *Procedia Engineering*. **100** 482-487.
[2] Kuczyk M, Jędrzejewski P, Załuski P 2022 Stress analysis of suspended rail vehicle bogie. *Rail Vehicles/Pojazdy Szynowe*. **1-2** 20-29.
[3] Abdurasulov S, Zayniddinov N, Yusufov A and Jamilov S 2023 Analysis of stress-strain state of bogie frame of PE2U and PE2M industrial traction unit. *E3S Web of Conferences*. **401** 1-9.
[4] Seo J W, Hur H M, Jun H K, Kwon S J, and Lee D H 2017 Fatigue design evaluation of railway bogie with full-scale fatigue test. *Advances in Materials Science and Engineering*. **2017(7)** 1-11.
[5] An Q, Zhao H, Li P and Fu M 2019 Fatigue strength analysis of bogie frames under random loads. *Advances in Mechanical Engineering*. **11(9)**1-13.
[6] Lutema M E and Mohammedseid A 2022 Fatigue strength analysis of the bogie frame for the CRH2 high-speed train in consideration of uncertainty parameters. *International Journal of Engineering and Applied Sciences*. **9(12)** 1-10.
[7] EUROPEAN STANDARD NORME EUROPÉENNE EUROPÄISCHE NORM, EN13749: 2011:E, Railway applications - Wheelsets and bogies - Method of specifying the structural requirements of bogie frames, CEN, 2011.
[8] EUROPEAN STANDARD NORME EUROPÉENNE EUROPÄISCHE NORM, EN15663: 2011:E, Railway applications - Definition of vehicle reference masses, CEN, 2009.
[9] Sapkal A B and Sirsikar S S 2017 Static and fatigue strength analysis of bogie frame. *National Conference in Applied Sciences* 2016, Mumbai, 129-134.
[10] Fang K T and Wang Y 1994 *Number-Theoretic Methods in Statistics*. Chapman & Hall, London.
[11] Cheng Y C, Yeh H C and Lee C K 2022 Multi-objective optimization of the honeycomb core in a honeycomb structure using uniform design and grey relational analysis. *Engineering Optimization*. **54(2)** 286-304.

Journal of Physics: Conference Series **2793** (2024) 012011 doi:10.1088/1742-6596/2793/1/012011

Preliminary analysis for a new sub-type III Solar Radio Burst (Type VI) on 11th July 2023

N.H.M. Bokhari[1,2], Z.S. Hamidi[1,2*] and N.N. M. Shariff[1,3]

[1] Faculty of Applied Sciences, Universiti Teknologi MARA, 40450 Shah Alam, Selangor, Malaysia
[2] Institute of Science, Universiti Teknologi MARA, 40450 Shah Alam, Selangor, Malaysia
[3] Academy of Contemporary Islamic Studies, Universiti Teknologi MARA, 40450 Shah Alam, Selangor, Malaysia

*Corresponding author's email: zetysh@uitm.edu.my

Abstract. We focus on the subcategory Solar Radio Burst Type III (Type VI) events detected on 11th July 2023, exhibiting an extended duration lasting almost 24 hours. Utilizing archival data from the e-CALLISTO network, we analyze the dynamical properties with emphasis on the 11:17-11:30 UTC interval due to their prominent features. The six selected stations recorded a series of well-defined SRB type III features with various intensities. Drift rates were calculated and produced the average absolute value of 1.4 MHz/s for a series burst and 8.2 MHz/s for a single burst, aligning with previous SRB type III drift rate ranges. Solar parameters on that day indicate a quite high range solar wind of 329.9 km/s, high solar coronal density of 6.07 protons/cm^3, also high solar flux units in the radio sun measurement correlating with the abundance of sunspots numbers and an elevated peak in solar flare intensity on that month. Direct associations between SRB type VI and solar flares of class C4.5 were observed, while no direct links to Coronal Mass Ejections (CMEs) were identified within that timeframe. This research underscores the importance of understanding SRB type VI characteristics and their associations with solar activities for space weather forecasting.

1. Introduction

Our Sun presently finds itself in the main sequence phase of the stellar evolution, characterized by stable magnetic activity and relatively low levels of hot plasma, X-ray emission and high energy particles [1]. Currently, it is undergoing the 25th solar cycle, a recurring 11-year period of solar activities although with a slight variation of around 14 months [2]. The peak of Solar Cycle 25 is expected to occur between August 2023 and July 2024 [3].

Solar Radio Bursts (SRBs) are brief periods of enhanced radio emission from the Sun. It is an arrangement of a frequency space that varies with time [4] and was the initial phenomenon recognized in the discipline of radio astronomy. The early observation was in 1944 by J.S Hey when he discovered radio emission of the Sun [5]. The solar radio frequency range, covering 70 MHz to 2.2 GHz, primarily facilitates the identification of the majority of SRBs, with a focus on the lower frequency spectrum, typically below 200 MHz [6].

SRBs are frequently associated with solar activities such as solar flares and Coronal Mass Ejection (CME). Solar flares are energetic events on the Sun that release energy from current-carrying magnetic fields [7]. It coincides with solar burst events particularly great solar flares of class X and M producing major storms [8]. Meanwhile, Coronal Mass Ejections (CMEs) are massive eruptions of plasma from the Sun, driven by magnetic forces, that can cause geomagnetic disturbances on Earth [9] which include communication & navigation disruptions, satellite interference and power grid issues.

Although there are well-known five categories of SRBs [10], [11], [12] depending on the frequencies since the 1960s there are still some types related to solar magnetic reconnection that are not detectable or recognizable due to the variability of the Sun [13]. However, in this paper, we will examine a sub-type III burst (SRB type VI) in more depth, which has never been done before. Solar radio burst type VI is a subcategory from type III, which is a series of type III bursts (fast drift bursts) over a period of 10 minutes or more, with no period longer than 30 minutes without activity [14]. Consequently, we hypothesize that SRB VI will affect the space weather and climate change on Earth as seen during SRB type III and affects more due to its continuous burst in long duration. The ongoing solar cycle is shown to have a lot of SRB type VI listed and identified by Monstein [15] from multiple radio instruments around the world.

As SRB type VI is a part of SRB type III, the SRB type III research has been ongoing for over half a century introduced by Wild [16]. SRBs type III are a signature of propagating beams of non-thermal electrons in the solar atmosphere and the solar system, its observational properties can help to identify attributes of electron beams and the ambient background plasma [17]. It has been characterized by rapid fast drift, short duration and large bandwidth [18]. High drift rate values are seen in SRB type III compared to others [6]. It can be commonly found in meter wavelength in the frequency range of 10 MHz to 500 MHz and the duration can exceed 30 minutes [19]. Statistically, it exhibits a clear maximum of their median radio flux density around 2 MHz in the metric to hectometric wavelength range [20]. SRB type III also is an indicator of the starting point of multiple magnetic reconnection [21], [22] where solar flares are associated [23] which is caused by the motion of plasma and other particles through the convection mechanism inside the Sun [24]. It has been emphasized that the importance of studying SRB type III due to its potential to offer insights into the energy release mechanism in solar flares, given its emission at decametric wavelength and relationship to the active region magnetic field [25], while this implies the same for SRB VI even more because it occurs in longer duration, this new subtype occurs frequently recent and will impact current solar cycle 25 which is climbing to its peak. In this paper, we will focus on analyzing the dynamical and kinematic properties of the SRBT VI using the E-CALLISTO archival data.

2. Methodology

Solar Radio Bursts type VI were analyzed thoroughly with the resources of the list of identified SRB type VI events that are retrieved from the e-CALLISTO open source website for Compound Astronomical Low-cost Low-frequency Instrument for Spectroscopy in Transportable Observatories (CALLISTO) [15]. It offers 24 hours of monitoring of the Sun from a worldwide initiative since 2007 [26]. One of its contributors, which we used in this paper, the Arecibo Observatory located in Puerto Rico, United States used the Long Wavelength Array (LWA) antennas for the SRB detections, and its utilization has proven to be quite a strong and relevant detection for advancing solar astrophysics [27]. The other five stations use a log periodic dipole antenna (LPDA) that has also proven to enhance the SRB research [28] in Germany and Glasgow, another LWA antenna in Swiss Muhen and Swiss Landschlacht and a simple Windom antenna in Norway. Some notable constraint worth mentioning is that most of the CALLISTO-based instruments have limitations in either frequency or time resolution, or they may lack sensitivity to detect certain types of faint radio emissions and this is primarily due to the antennas' low effective area and the noisy environments where the stations are located [29].

Journal of Physics: Conference Series **2793** (2024) 012011 doi:10.1088/1742-6596/2793/1/012011

As a part of the characterization of the solar radio burst, we calculate the drift rate using the equation

$$Drift\ rate\ \left(\frac{df}{dt}\right) = \frac{(f_e - f_s)}{(t_e - t_s)} \left(\frac{MHz}{s}\right) \quad (1)$$

where f_e is the final frequency, f_s is the starting frequency, t_e is the end time and t_s is the starting time of the bursts. Through the computation of the drift rate, we can determine the duration of a burst, distinguishing between those with extended or brief lifespans. For accurate determination of the starting and final frequencies, SAO Image D9 software is employed as a raw data visualization tool. To enhance the visibility of fainter data, the contour level parameter is adjusted to delineate the boundaries of the starting and final frequencies.

3. Results & discussion

On 11th July 2023, despite the longest SRB duration reported that nearly hit the 24-hour mark, there are three distinct significant detections of SRB type VI were highlighted, each characterized by a relatively shorter duration, specifically from 10:46 to 11:00 UTC, 11:17 to 11:30 UTC, and 17:30 to 17:44 UTC. It is worth noting that the first period may have a slightly extended duration than what was initially reported. However, our detailed analysis will concentrate on the time interval from 11:17 to 11:30 UTC due to its prominent and noteworthy features. It was identified across 27 stations globally, but we will specifically emphasize six stations that consistently detected the other two events on that day which are Arecibo Observatory, Germany-DLR, Glasgow, Norway-Egersund, Swiss-Landschlacht and Swiss-Muhen. Figure 1 shows the SRB type VI observed by Arecibo Observatory, displaying a series of highly defined SRB type III vertical features in the timeframe of around 13 minutes. It shows around 2 minutes and 3 minutes of inactivity in between the series of bursts. Some small and fainter features are also seen in the background contributing to the continuation of the SRB although it is not taken into account when calculating the drift rate. Other station also exhibits the same structure but only differ in terms of intensity. Although, Glasgow station data had some radio interference frequency that affected the drift calculation but we have taken into account the second-best level of intensity that did not interfere with the data.

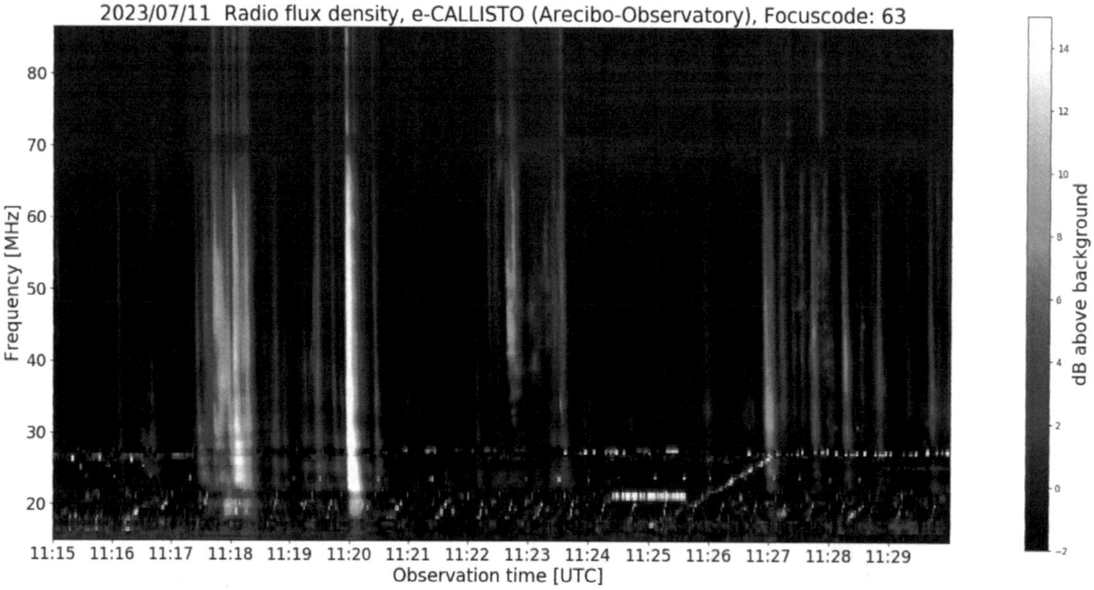

Figure 1. Solar Radio Burst Type VI observed by Arecibo Observatory on 11th July 2023

Journal of Physics: Conference Series **2793** (2024) 012011 doi:10.1088/1742-6596/2793/1/012011

Using Equation (1), we calculate the drift rate separately for a series of bursts and a well-isolated single burst as visibly seen at 11:18 UTC and 11:20 UTC respectively. Figure 2 shows the result for all six stations and the absolute values are labeled for each series and single burst. The range absolute drift rate for the series drift rate is ~0.9 to 2.1 MHz/s with an average of 1.4 MHz/s. On the other hand, the range absolute drift rate for the single burst is ~6.5 to 10.6 MHz/s with an average of 8.2 MHz/s. Due to the fact that SRB type VI comprises a series of type III, its drift rates align with other ranges of SRB type III drift rates, for series bursts with the range of absolute value of ~0.5 to 6 MHz/s [30] while for the well-isolated single burst that usually ranges from ~2 to 16 MHz/s with the median value of 6.94 MHz/s [31]. Moreover, we observed that all the data exhibited the explosion value of less than 14 dB, consistent with previous findings for SRB type III [32].

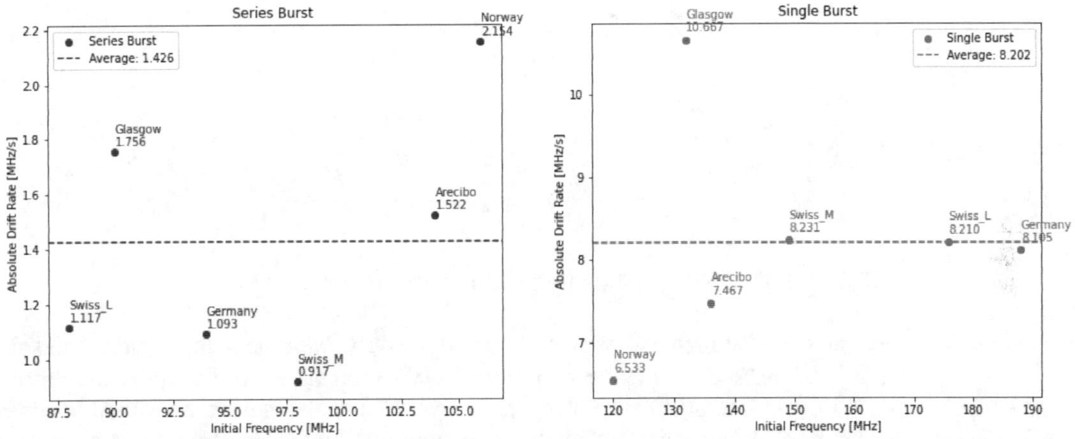

Figure 2. Drift rate for series and single bursts on 11th July 2023

Table 1 presents relevant solar parameters on 11th July 2023. The solar wind is reported at 11:17 UTC at a speed of 329.9 km/s, indicating a relatively quite high solar wind while the highest peak on that day is 442.9 km/s which occurs at 1:19 UTC. A higher solar wind speed enhances the likelihood of improved geomagnetic conditions. The solar density in the corona is measured at 6.07 protons/cm^3 at 11:18 UTC, also considered as in the high range, while the highest peak on that day is 9.51 protons/cm^3 at 11.44 UTC. This density is associated with auroral activity in Earth's polar regions, where higher density increases the chances of such occurrences. The radio sun measurement at 10.7 cm records 214 solar flux units (s.f.u), correlating with a sunspot number of 227 which also positioned it in the high ranges for the parameters. The planetary Kp index is finalized at 3-, indicating an unsettled condition, but remains below storm levels.

Table 1. The solar parameter on 11th July 2023 (Credit: NOAA SWPC)

Parameter	Value
Solar Wind (11:17 UTC)	329.9 km/s
Solar Density (11:18 UTC)	6.07 p/cm^3
Radio Sun (10.7 cm)	214 sfu
Sunspot Number	227
Finalized Kp Index	3-

Journal of Physics: Conference Series **2793** (2024) 012011 doi:10.1088/1742-6596/2793/1/012011

Subsequently, we investigate solar observations in the X-ray regime, revealing a direct association between SRB type VI and a solar flare recorded in the GOES X-ray data, classified as C4.5 that valued as 4.5×10^{-6} watts/m^2 from 11:13 UTC to 11:35 UTC, which peaks at 11:23 UTC, as shown in Figure 3. This particular solar flare originates from AR 13363 which at the time consisted of 21 sunspots with a size of 780 MH and magnetic classification of β which means a bipolar region with simple division between the polarities as seen in Figure 4.

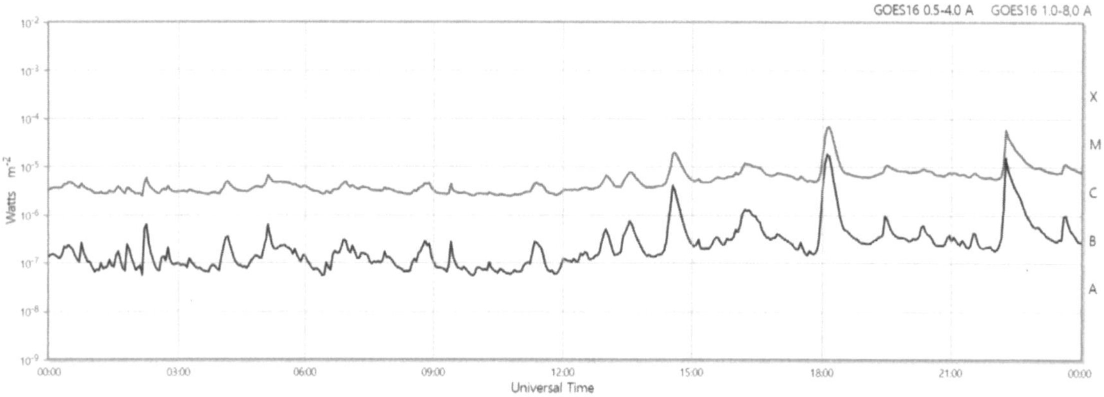

Figure 3. GOES X-ray flux on 11th July 2023 (Credit: KSWC)

On this particular day, the solar flare intensity reached its peak for July, with the highest maximum solar flare recorded at M6.83 and marked the highest count of M-class solar flares for the entire month of 6 counts while 8 counts for C-class solar flares. However, we have to note that solar flare M6.83 occurred only 7 minutes after the third period of SRB type VI which ends at 17:44 UTC. Another M class solar flare, M1.1 also occurred not directly within the timeframe but 55 minutes before that particular SRB that started at 17:30 UTC. In terms of the CMEs, it is recorded to have five CME events but not directly within the timeframe of the SRB type VI series on that day.

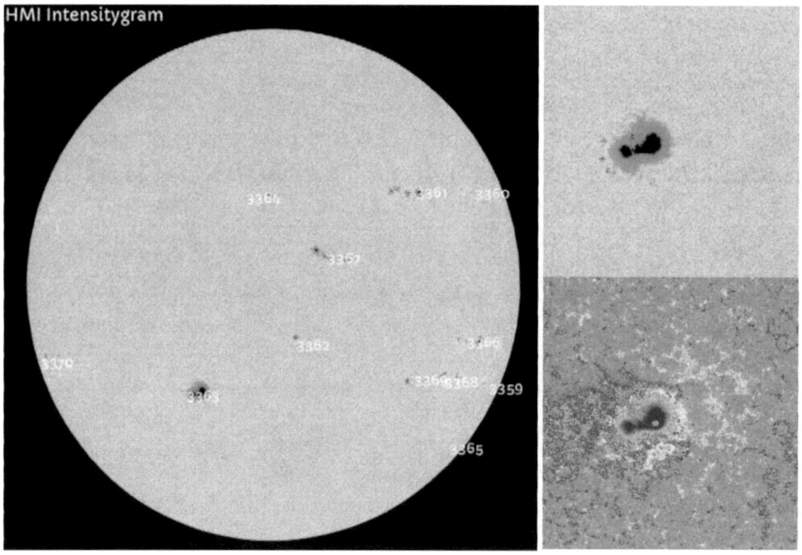

Figure 4. (Left) Active Regions on 11th July 2023 (Right) AR13363 from SDO HMIIF (above) and SDO HMIBC (below) (Credit: SDO/SOHO/HMI)

4. Conclusion

In summary, SRB type VI was consistently observed and reported throughout 2023 via the CALLISTO network, with a notable instance of prolonged solar radio bursts on 11th July 2023, spanning nearly a full day. The six observing stations that recorded SRB type VI on this date demonstrated similar drift rate ranges with an average absolute value of 1.4 MHz/s for a series burst and 8.2 MHz/s for a single burst, aligning with previous findings. We found a direct association between C4.5 solar flares with one of the specified periods of SRB type VI, while no direct association with CMEs was identified within the selected timeframe of the prominent solar radio bursts type VI. Further research is needed to explore correlations between solar flares, CMEs, and SRB type VI on a larger scale, especially with the impending peak of solar cycle 25, which promises increased observations of solar activities.

Acknowledgements

We express our gratitude to the CALLISTO network, LASCO, SDO/AIA, NOAA, and SWPC for providing accessible online data. Special thanks to C. Monstein, the PI for the e-Callisto project, for his diligent efforts in visually inspecting and reporting the SRB event list from extensive archival data, ensuring accessibility for all. We also recognize the support received through grant FRGS/1/2023/STG07/UiTM102/11 and from the Ministry of Higher Education, Malaysia.

References

[1] M. Güdel, "The sun in time: From PMS to main sequence," *EAS Publications Series*, vol. 9, pp. 339–339, 2003, doi: 10.1051/eas:2003119.

[2] D. H. Hathaway, "The solar cycle," *Living Rev Sol Phys*, vol. 12, pp. 1–81, Feb. 2015, doi: 10.1007/lrsp-2015-4.

[3] A. Hu and A. Rodriguez, "A hybrid CNN-LSTM model for predicting solar cycle 25," *Journal of Student Research*, vol. 12, no. 1, 2023.

[4] A. Hanslmeier and M. Messerotti, *Motions in the solar atmosphere*, vol. 239. Springer Science & Business Media, 1999.

[5] J. S. Hey, S. J. Parsons, and J. W. Phillips, "Some characteristics of solar radio emissions," *Mon Not R Astron Soc*, vol. 108, no. 5, pp. 354–371, 1948.

[6] J. V. Wijesekera, K. P. S. C. Jayaratne, and J. Adassuriya, "Analysis of type II and type III solar radio bursts," *J Phys Conf Ser*, vol. 1005, no. 1, p. 012046, May 2018, doi: 10.1088/1742-6596/1005/1/012046.

[7] A. G. Emslie, "Overview of solar flares," *The Sun: A Laboratory for Astrophysics* , pp. 465–474, 1992.

[8] Z. S. Hamidi, N. Ramli, and N. N. M. Shariff, "Statistical analysis of solar radio burst type III and type IV with relation to solar activities," *J Phys Conf Ser*, vol. 1152, no. 1, p. 012018, Mar. 2019, doi: 10.1088/1742-6596/1152/1/012018.

[9] P. J. Cargill and L. K. Harra, "Coronal mass ejection," *Handbook of the Solar-Terrestrial Environment*, p. 118, 2007.

[10] M. R. Kundu, "Solar radio astronomy," *New York: Interscience Publication*, 1965.

[11] D. J. McLean and N. R. Labrum, "Solar radiophysics: Studies of emission from the sun at metre wavelengths," *Solar Radiophysics: Studies of Emission from the Sun at Metre Wavelengths*, 1985.

[12] A. Benz, "Kinetic plasma and particle beams," *Plasma Astrophysics: Kinetic Processes in Solar and Stellar Coronae*, pp. 89–114, 1993.

[13] N. Husien *et al.*, "Solar radio bursts detected by CALLISTO system and their related events," in *International Conference on Industrial Engineering, Management Science and Application (ICIMSA)*, IEEE, 2016, pp. 148–152.

[14] "Explanation of the information in SWPC solar and geophysical event reports." Accessed: Nov. 23, 2023. [Online]. Available: https://www.e-callisto.org/eventSummaryExplanation.pdf

[15] C. Monstein, A. Csillaghy, and A. O. Benz, "CALLISTO solar spectrogram FITS files," *International Space Weather Initiative (ISWI)*. 2023. doi: 10.48322/pmwd-mk15.

[16] J. P. Wild, S. F. Smerd, and Weiss A. A., "Solar bursts," *Annu Rev Astron Astrophys*, vol. 1, no. 1, pp. 291–366, 1963.

[17] H. A. S. Reid and H. Ratcliffe, "A review of solar type III radio bursts," *Res Astron Astrophys*, vol. 14, no. 7, p. 773, Apr. 2014, doi: 10.1088/1674-4527/14/7/003.

[18] N. A. M. Norsham and Z. S. Hamidi, "The relationship between solar radio burst type III and geomagnetic storm on the earth that occurred on 4th October 2017," *J Phys Conf Ser*, vol. 1152, no. 1, p. 012019, Mar. 2019, doi: 10.1088/1742-6596/1152/1/012019.

[19] Z. S. Hamidi, N. N. M. Shariff, Z. A. Ibrahim, and N. Ramli, "Investigation of the statistical properties of solar radio burst type II and III," in *International Conference on Space Science and Communication (IconSpace)*, 2015, pp. 335–338.

[20] K. Sasikumar Raja *et al.*, "Spectral analysis of solar radio type III bursts from 20 kHz to 410 MHz," *Astrophys J*, vol. 924, no. 2, p. 58, Jan. 2022, doi: 10.3847/1538-4357/ac34ed.

[21] Z. S. Hamidi and N. N. M. Shariff, "The analysis of solar radio burst type III associated with solar activities due to magnetic reconnection from 2012-2017," *J Phys Conf Ser*, vol. 1593, no. 1, p. 012025, Aug. 2020, doi: 10.1088/1742-6596/1593/1/012025.

[22] N. Mohamad Ansor, Z. S. Hamidi, and N. N. M. Shariff, "Application of Solar Radio Bursts Observation in Detecting Magnetic Reconnection Phenomenon," in *Journal of Physics: Conference Series*, IOP Publishing Ltd, Aug. 2020. doi: 10.1088/1742-6596/1593/1/012024.

[23] N. M. Ansor, Z. S. Hamidi, and N. N. M. Shariff, "Characterization of type III and IV solar radio bursts from e-CALLISTO," *J Phys Conf Ser*, vol. 1349, no. 1, p. 012063, Dec. 2019, doi: 10.1088/1742-6596/1349/1/012063.

[24] Z. S. Hamidi *et al.*, "Magnetic reconnection of solar flare detected by solar radio burst type III," *J Phys Conf Ser*, vol. 539, no. 1, p. 012006, 2014, doi: 10.1088/1742-6596/539/1/012006.

[25] Z. S. Hamidi, U. F. S. U. Ibrahim, Z. Z. Abidin, Z. A. Ibrahim, and N. N. M. Shariff, "Theoretical review of solar radio burst III (SRBT III) associated with of solar flare phenomena," *International Journal of Fundamental Physical Sciences*, vol. 3, no. 2, pp. 20–23, Jun. 2013, doi: 10.14331/ijfps.2013.330048.

[26] A. O. Benz *et al.*, "A world-wide net of solar radio spectrometers e-CALLISTO," in *Earth, Moon, and Planets*, 2009, pp. 277–285. doi: 10.3929/ethz-b-000157338.

[27] G. Carson, J. E. Kooi, J. F. Helmboldt, B. B. Markowski, D. J. Bonanno, and B. C. Hicks, "DLITE—An inexpensive, deployable interferometer for solar radio burst observations," *Frontiers in Astronomy and Space Sciences*, vol. 9, p. 1026455, 2022.

[28] N. Mohamad Ansor, Z. S. Hamidi, and N. N. M. Shariff, "The implementation of Log Periodic Dipole Antenna (LPDA) and CALLISTO system on solar radio bursts observation," *J Phys Conf Ser*, vol. 1768, no. 1, Feb. 2021, doi: 10.1088/1742-6596/1768/1/012011.

[29] "E-CALLISTO antenna." Accessed: Jan. 28, 2024. [Online]. Available: https://www.assa.org.au/media/74579/assa-antenna.pdf

[30] W. Z. A. Wan Mokhtar, Z. S. Hamidi, Z. Z. Abidin, and Z. A. Ibrahim, "Drift frequency of kilometric radio burst type III with solar wind speed and density," *J Phys Conf Ser*, vol. 1152, no. 1, p. 012032, Mar. 2019, doi: 10.1088/1742-6596/1152/1/012032.

[31] P. J. Zhang, C. B. Wang, and L. Ye, "A type III radio burst automatic analysis system and statistic results for a half solar cycle with Nançay Decameter Array data," *Astron Astrophys*, vol. 618, Oct. 2018, doi: 10.1051/0004-6361/201833260.

[32] Z. S. Hamidi, L. Chen Chen, and Z. K. D. Nurlisman, "Moderate radio flux density of single and group III bursts from AR2887 Active Region on 26th October 2021," in *Journal of Physics: Conference Series*, Institute of Physics, 2022. doi: 10.1088/1742-6596/2287/1/012033.

Journal of Physics: Conference Series **2793** (2024) 012012 doi:10.1088/1742-6596/2793/1/012012

Linear intersection pairs of negacyclic codes and their applications in construction of entanglement-assisted quantum error-correcting codes

Somphong Jitman

Department of Mathematics, Faculty of Science, Silpakorn University, Nakhon Pathom 73000, Thailand

Corresponding author's email: sjitman@gmail.com

Abstract. Linear intersection pairs of linear codes have become of interest and continuously studied due to their nice algebraic properties and wide applications. In this article, we focus on linear intersection pairs of negacyclic codes over finite fields and their applications. General characterization and algebraic properties of such pairs are given in terms of their generator polynomials. For $s \in \{0,1\}$, explicit constructions of linear s-intersection pairs and linear s-complementary pairs of negacyclic codes are presented. As applications, constructions of entanglement-assisted quantum error-correcting codes are discussed together with illustrative examples.

1. Introduction

Families of cyclic and negacyclic codes over finite fields have been of interest since they have nice algebraic structures and can be applied in various fields. The algebraic structures of such codes are given though the ideals in $F_q[x]/<x^n-1>$ and $F_q[x]/<x^n+1>$ (see [1]). In applications, such algebraic structured codes can be easily implemented in shift registers [2]. Linear complementary pairs (LCPs) of linear codes over finite fields have been introduced in [3] and shown to have applications in cryptography. In [3] and [4], LCPs have been applied in in counter passive and active side-channel analysis attacks on embedded crypto systems. In [3] and [4], several constructions of LCPs of linear codes were given as well.

In [5], the author formulated the notion of a linear intersection pair of linear codes. Subsequently, their applications in constructions of entanglement-assisted quantum error-correcting codes (EAQECCs) have been presented as well. For an integer $s \geq 0$, a linear s-intersection pair of linear codes is defined to be linear codes C and D of the same length n over F_q whose intersection has dimension s. This can be viewed as a generalization of various concepts in coding theory such as complementary dual codes, self-orthogonal codes, hulls, and LCPs.

Recently, linear s-intersection pairs of cyclic codes were studied in [6] and linear 1-complementary pairs of negacyclic codes were studied in [7]. In this article, we extend some concepts in [6] and [7] to s-intersection pairs and s-complementary pairs of negacyclic codes over finite fields. Such pairs have different parameters compare to [6] and they are optimal in some cases. Moreover, such pairs have more algebraic structures than the pairs of linear codes in [5] and they can be contracted directly using ideals in polynomial rings. Precisely, we focus on linear s-intersection pairs and linear s-complementary pairs of negacyclic codes over finite fields. General characterization of such pairs of negacyclic codes is given in terms of their generator polynomials. For $s \in \{0,1\}$, explicit constructions of linear s-intersection pairs and linear s-complementary pairs of negacyclic codes are given as well as applications in construction of EAQECCs. Some illustrative examples of linear intersection pairs derived from negacyclic codes are provided. Some of them have optimal parameters. Examples of EAQECCs and EAQECCs with positive net-rate are presented as well.

After this introductory section, basic concepts and results on codes and linear intersection pairs are recalled in Section 2. Characterizations of linear s-intersection pairs and linear s-complementary pairs of negacyclic codes are presented in Section 3 together with constructions of two negacyclic codes whose intersecting dimension is 0 or 1. Applications in constructions of EAQECCs are provided in Section 4 as well as illustrative examples. Summary and discussion are provided in Section 5.

2. Preliminaries

Basic concepts and properties of linear codes, negacyclic codes, linear intersection pairs, linear complementary pairs are recalled. For more details, the reader may refer to [1], [2], [4], [5], and [7].

2.1. Linear intersection pairs of codes

Let q be a prime power and let n be a positive integer. Let F_q be a finite field of q elements. A linear code of length n is defined to be a subspace of $(F_q)^n$. An $[n, k]_q$ code refers to a linear code C of length n over F_q whose dimension dim(C) is k. An $[n, k]_q$ code C is called an $[n, k, d]_q$ code if the minimum hamming weight of C is d. We denote by $C^{\perp}$ the dual code of C under the Euclidean inner product.

Linear codes C_1 and C_2 of length n over F_q form a linear complementary pair (LCP) if $C_1 \cap C_2=\{\mathbf{0}\}$ and $C_1+C_2=(F_q)^n$ (see [4]). For a non-negative integer s, C_1 and C_2 are said to be a linear s-intersection pair if $\dim(C_1 \cap C_2)=s$ (see, [5]). A linear s-intersection pair C_1 and C_2 is called a linear s-complementary pair if $C_1+C_2=(F_q)^n$ (cf. [7]). Clearly, a linear 0-complementary pair is a LCP.

2.2. Negacyclic codes over finite fields

A linear code C of length n over F_q is called a negacyclic code if $(c_{n-1}, c_0, \ldots, c_{n-2}) \in C$ for all $(c_0, c_1, \ldots, c_{n-1}) \in C$ (see [7]). It is well known that every negacyclic code of length n over F_q can be represented by a principal ideal in $F_q[x]/\langle x^n+1\rangle$ whose generator is a unique monic divisor of x^n+1 of minimal degree. Such a monic polynomial is called the generator polynomial of the negacyclic code C.

Properties of negacyclic codes over finite fields are given in terms of their generator polynomials (see, for example, [1] and [7]).

Theorem 2.1: Let C be a negacyclic code of length n over F_q with generator polynomial g(x) and let k be an integer such that $0 \le k \le n$. Then C has dimension k if and only if deg(g(x))=n-k.

Theorem 2.2: Let C_1 and C_2 be negacyclic codes of length n over F_q with generator polynomials $g_1(x)$ and $g_2(x)$, respectively. Then $C_1 \cap C_2$ is a negacyclic code of length n generated by $\mathrm{lcm}(g_1(x), g_2(x))$ and C_1+C_2 is a negacyclic code of length n generated by $\gcd(g_1(x), g_2(x))$.

3. Linear intersection pairs of negacyclic codes over finite fields

Characterizations of a linear s-intersection pair and a linear s-complementary pair of negacyclic codes are presented. For $s \in \{0,1\}$, explicit constructions of such pairs are given as well.

3.1. Characterization of linear intersection pairs and linear complementary pairs of negacyclic codes

General characterizations of a linear s-intersection pair and a linear s-complementary pair of negacyclic codes are given as follows.

Theorem 3.1: Let F_q be a finite field and let n be a positive integer co-prime to q. Let $s \ge 0$ be an integer. Let C_1 and C_2 be negacyclic codes of length n over F_q with generator polynomials $g_1(x)$ and $g_2(x)$, respectively. Then the following statements hold.

(1) C_1 and C_2 form a linear s-intersection pair if and only if $\deg(\mathrm{lcm}(g_1(x), g_2(x)))=n-s$.

(2) C_1 and C_2 form a linear s-complementary pair if and only if $\deg(\mathrm{lcm}(g_1(x), g_2(x)))=n-s$ and $\gcd(g_1(x),g_2(x))=1$.

Proof. First, we prove 1). Assume that C_1 and C_2 form a linear s-intersection pair. Then $\dim(C_1 \cap C_2)=s$. By Theorem 2.2, $C_1 \cap C_2$ is generated by $\mathrm{lcm}(g_1(x), g_2(x))$. Hence, $\deg(\mathrm{lcm}(g_1(x), g_2(x)))=n-s$ by

Journal of Physics: Conference Series **2793** (2024) 012012 doi:10.1088/1742-6596/2793/1/012012

Theorem 2.1. Conversely, assume that $\deg(\mathrm{lcm}(g_1(x), g_2(x)))=n-s$. By Theorem 2.2, $C_1 \cap C_2$ is negacyclic whose generator polynomial is $\mathrm{lcm}(g_1(x), g_2(x))$. Since $\deg(\mathrm{lcm}(g_1(x), g_2(x)))=n-s$, it follows that $\dim(C_1 \cap C_2)=s$ by Theorem 2.1. Hence, C_1 and C_2 form a linear s-intersection pair.

Next, we note that $C_1+C_2=(F_q)^n$ if and only if C_1+C_2 is generated by $1= \gcd(g_1(x), g_2(x))$. Hence, 2) follows from the previous discussion and 1).

Some illustrative examples of a linear s-intersection pair are given.

Example 3.1: Let C_1 and C_1 be $[6, 3]_5$ and $[6, 3]_5$ negacyclic codes of length 6 over F_5 with generator polynomials $g_1(x)=x^3 + 2$ and $g_2(x)=x^3 + x^2 + 3x + 2$, respectively. Since $\mathrm{lcm}(g_1(x), g_2(x))= x^5 + 3x^4 + 4x^3 + 2x^2 + x + 3$, we have $\deg(\mathrm{lcm}(g_1(x), g_2(x)))=5=6-1$ which implies that C_1 and C_1 form a linear 1-intersection pair. Since $\gcd(g_1(x), g_2(x))= x+3 \neq 1$, C_1 and C_1 are not linear 1-complementary pair.

Example 3.2: Let C_1 and C_1 be $[6, 3]_5$ and $[6, 4]_5$ negacyclic codes of length 6 over F_5 with generator polynomials $g_1(x)=x^2 + 2x + 4$ and $g_2(x)=x^3 + x^2 + 3x + 2$, respectively. Since $\mathrm{lcm}(g_1(x), g_2(x))= x^5 + 3x^4 + 4x^3 + 2x^2 + x + 3$, we have $\deg(\mathrm{lcm}(g_1(x), g_2(x)))=5=6-1$ which implies that C_1 and C_1 form a linear 1-intersection pair. Since $\gcd(g_1(x), g_2(x))= 1$, C_1 and C_1 are also a linear 1-complementary pair.

3.2. Constructions of linear intersection pairs negacyclic codes

A general construction for linear s-intersection pairs of negacyclic codes is given.

Theorem 3.2: Let F_q be a finite field and let n be a positive integer such that $\gcd(n,q)=1$. Let $s \geq 0$ be an integer and let $e(x)$ be a monic divisor of x^n+1 of degree s. If $a(x)$ and $b(x)$ are monic polynomials such that $b(x)|a(x)| \frac{x^n+1}{e(x)}$ over F_q, then the negacyclic codes with generator polynomials $a(x)$ and $c(x):=\frac{(x^n+1)b(x)}{e(x)a(x)}$ form a linear s-intersection pair.

Proof. From the definition, we have $\mathrm{lcm}(a(x), c(x))=\frac{x^n+1}{e(x)}$ which implies that $\deg(\mathrm{lcm}(a(x), c(x)))=n-s$. It follows that the negacyclic codes with generator polynomials $a(x)$ and $c(x)$ form a linear s-intersection pair by (1) of Theorem 3.1.

By setting $e(x)=1$ in Theorem 3.2, a linear 0-intersection pair of negacyclic codes can be derived in the following corollary.

Corollary 3.3: Let F_q be a finite field and let n be a positive integer such that $\gcd(n,q)=1$. Let $a(x)$ and $b(x)$ be monic polynomials such that $b(x)|a(x)|(x^n+1)$ over F_q. Then the negacyclic codes with generator polynomials $a(x)$ and $c(x):=\frac{(x^n+1)b(x)}{a(x)}$ form a linear 0-intersection pair.

Clearly, $(x+1)|(x^n+1)$ for all odd positive integer n. By setting $e(x)=x+1$ in Theorem 3.2, a construction of linear 1-intersection pairs of negacyclic codes is presented as follows.

Corollary 3.4: Let F_q be a finite field and let n be an odd positive integer such that $\gcd(n,q)=1$. Let $a(x)$ and $b(x)$ be monic polynomials such that $b(x)|a(x)| \frac{x^n+1}{x+1}$ over F_q. Then the negacyclic codes with generator polynomials $a(x)$ and $c(x):=\frac{(x^n+1)b(x)}{(x+1)a(x)}$ form a linear 1-intersection pair.

Constructions of linear s-complementary pairs of negacyclic codes are now given.

Theorem 3.5: Let F_q be a finite field and let n be a positive integer such that $\gcd(n,q)=1$. Let $e(x)$ be a monic divisor of x^n+1 of degree s. If $a(x)$ is a monic polynomial such that $a(x)| \frac{x^n+1}{e(x)}$ over F_q, then the negacyclic codes with generator polynomials $a(x)$ and $c(x):=\frac{(x^n+1)}{e(x)a(x)}$ form a linear s-complementary pair.

Proof. We note that $\mathrm{lcm}(a(x), c(x))=\frac{x^n+1}{e(x)}$ which implies that $\deg(\mathrm{lcm}(a(x), c(x)))=n-s$. Since $\gcd(a(x), c(x))=1$, the negacyclic codes with generator polynomials $a(x)$ and $c(x)$ form a linear s-complementary pair by (2) of Theorem 3.1.

By setting $e(x)=1$ in Theorem 3.5, the next corollary follows.

Journal of Physics: Conference Series **2793** (2024) 012012 doi:10.1088/1742-6596/2793/1/012012

Corollary 3.6: Let F_q be a finite field and let n be a positive integer such that gcd(n,q)=1. Let a(x) be monic polynomial such that $a(x)|(x^n+1)$ over F_q. Then the negacyclic codes with generator polynomials a(x) and $c(x):=\frac{x^n+1}{a(x)}$ are a linear 0-complementary pair (LCP).

By putting e(x)=x+1 in Theorem 3.5, a linear 1-complementary pairs of negacyclic codes is derived in the next corollary.

Corollary 3.7: Let F_q be a finite field and let n be an odd positive integer such that gcd(n,q)=1. Let a(x) be monic polynomial such that $a(x)|\frac{x^n+1}{x+1}$ over F_q. Then the negacyclic codes with generator polynomials a(x) and $c(x):=\frac{x^n+1}{(x+1)a(x)}$ form a linear 1-complementary pair.

In practice, linear intersection pairs of negacyclic codes can be constructed explicitly using Theorem 3.2 in the following steps: 1) Fix an integer s≥0. 2) Fix a monic divisor e(x) of x^n+1 (if exist). 3) Compute all monic divisors of $\frac{(x^n+1)}{e(x)}$. 4) For each monic divisor a(x) of $\frac{(x^n+1)}{e(x)}$, compute all monic divisors of a(x). 5) For each monic divisor b(x) of a(x), compute parameters of negacyclic codes with generator polynomials a(x) and $c(x):=\frac{(x^n+1)b(x)}{e(x)a(x)}$. The output codes form linear s-intersection pairs. In the same fashion, linear s-complementary pairs of negacyclic codes can be computed via Theorem 3.5.

Some illustrative examples of linear s-intersection pairs and linear s-complementary pairs of negacyclic codes over F_3 are given in Table 1. The negacyclic codes in Table 1 are optimal according to the database used in [8]. While the pairs in Table 1 are linear s-intersection pairs, the ones with * are also linear s-complementary pairs. For convenience, denote by $a_0a_1a_2\ldots a_r$ the polynomial $a_0+a_ax+a_2x^2+\cdots+a_rx^r$ over F_3. Using the notation as in Theorem 3.2, the negacyclic codes C_1 and C_2 are generated by a(x) and c(x), respectively.

Table 1. Linear s-intersection pairs and linear s-complementary pairs of negacyclic codes over F_3

e(x)	a(x)	b(x)	c(x)	C_1	C_2	s
1	1121	1	12221201011	$[13, 10, 3]_3$	$[13, 3, 9]_3$	0*
1	1021201	1	10122101	$[13, 7, 5]_3$	$[13, 6, 6]_3$	0*
1	1021201	1121	11010212221	$[13, 7, 5]_3$	$[13, 3, 9]_3$	0
1	1012010221	1	10211	$[13, 4, 7]_3$	$[13, 9, 3]_3$	0*
1	1012010221	1121	11110201	$[13, 4, 7]_3$	$[13, 6, 6]_3$	0
1	1012010221	1111211	11010212221	$[13, 4, 7]_3$	$[13, 3, 9]_3$	0
11	1121	1	1111021001	$[13, 10, 3]_3$	$[13, 4, 7]_3$	1*
11	1021201	1	1220221	$[13, 7, 5]_3$	$[13, 7, 5]_3$	1*
11	1021201	1121	1001201111	$[13, 7, 5]_3$	$[13, 4, 7]_3$	1
11	1012010221	1111211	1001201111	$[13, 4, 7]_3$	$[13, 4, 7]_3$	1
1201	1121	1	10201111	$[13, 10, 3]_3$	$[13, 6, 6]_3$	3*
1201	1121	1121	11102112101	$[13, 10, 3]_3$	$[13, 3, 9]_3$	3
1201	1021201	1	11201	$[13, 7, 5]_3$	$[13, 9, 3]_3$	3*
1201	12220021	1	1211	$[13, 6, 6]_3$	$[13, 10, 3]_3$	3*
1201	1012010221	1	11	$[13, 4, 7]_3$	$[13, 12, 2]_3$	3*
1201	1012010221	1121	12001	$[13, 4, 7]_3$	$[13, 9, 3]_3$	3
1201	1012010221	1111211	12220021	$[13, 4, 7]_3$	$[13, 6, 6]_3$	3

4. EAQECCs from linear s-intersection pairs of negacyclic codes over finite fields

Entanglement-assisted quantum error correcting codes (EAQECCs) were introduced in [9] and it has been shown that EAQECCs can be constructed from classical codes. Further, the performance of the resulting quantum codes can be determined by the performance of the underlying classical codes. Precisely, an $[[n,k,d;c]]_q$ EAQECC encodes k logical qudits into n physical qudits using c copies of maximally entangled states and its performance is measured by its rate $\frac{k}{n}$ and net rate $\frac{k-c}{n}$. When the net rate of an EAQECC is positive it is possible to obtain catalytic codes as shown in [8]. In [9], good entanglement-assisted quantum codes were constructed. A link between the number of maximally shared qubits required to construct an EAQECC from a classical code and the hull of the classical code were given. For more details on EAQECCs, please refer to [11], [12] and the references therein.

In [11], EAQECCs are constructed from linear intersection pair of classical linear codes.

Theorem 4.1 ([11, Proposition 4.2]): Let $s \geq 0$ be an integer and let C_1 and C_2 be a linear s-intersection pair of linear codes over F_q with parameters $[n, k_1, d_1]_q$ and $[n, k_2, d_2]_q$, respectively. Then there exists an $[[n, k_2-s, \min\{d_1^{\perp}, d_2\}; k_1-s]]_q$ EAQECC with $d_1^{\perp}$ is the minimum Hamming weight of $C_1^{\perp}$.

Based on Theorem 4.1, Theorem 3.2, Corollary 3.3, and Corollary 3.4, constructions of EAQECCs are given in terms of linear intersection pair of negacyclic codes as follows.

Theorem 4.2: Let F_q be a finite field and let n be a positive integer such that gcd(n,q)=1. Let e(x) be a monic divisor of x^n+1 of degree s. If a(x) and b(x) are monic polynomials such that $b(x)|a(x)|\frac{x^n+1}{e(x)}$, then there exists an $[[n, \deg(a(x))-\deg(b(x)), \min\{d_1, d_2\}; n-\deg(a(x))-s]]_q$ EAQECC, where d_1 and d_2 are the minimum Hamming weights of negacyclic codes generated by $\frac{x^n+1}{a(x)}$ and $c(x):=\frac{(x^n+1)b(x)}{e(x)a(x)}$, respectively.

Proof. Based on Theorems 3.2 and 4.1, the desired EAQECC can be constructed directly.

By setting e(x)=1 and e(x)=x+1 in Theorem 4.2, the next corollaries follow immediately.

Corollary 4.3: Let F_q be a finite field and let n be a positive integer such that gcd(n,q)=1. Let a(x) and b(x) be monic polynomials such that $b(x)|a(x)|(x^n+1)$. Then there exists an $[[n, \deg(a(x))-\deg(b(x)), \min\{d_1, d_2\}; n-\deg(a(x))]]_q$ EAQECC, where dis the minimum Hamming weight of the negacyclic code generated by $\frac{x^n+1}{a(x)}$.

Corollary 4.4: Let F_q be a finite field and let n be an odd positive integer such that gcd(n,q)=1. Let a(x) and b(x) be monic polynomials such that $b(x)|a(x)|\frac{x^n+1}{x+1}$. Then there exists an $[[n, \deg(a(x))-\deg(b(x)), \min\{d_1, d_2\}; n-\deg(a(x))-1]]_q$ EAQECC, where d_1 and d_2 are the minimum Hamming weights of negacyclic codes generated by $\frac{x^n+1}{a(x)}$ and $c(x):=\frac{(x^n+1)b(x)}{(x+1)a(x)}$, respectively.

4.1. Examples

In general, EAQECCs can be derived from linear s-intersection pairs of negacyclic codes using Theorem 4.2, Corollary 4.3, and Corollary 4.4. Some illustrative examples of EAQECCs over F_3 are presented in Table 2. The input negacyclic codes C_1 and C_2 with generator polynomials a(x) and c(x) are chosen from Table 1

Table 2. EAQECCs from linear intersection pairs of negacyclic codes over F_3

C_1	C_2	a(x)	c(x)	s	Q
$[13, 10, 3]_3$	$[13, 3, 9]_3$	1121	12221201011	0	$[[13, 3, 9; 10]]_3$
$[13, 7, 5]_3$	$[13, 6, 6]_3$	1021201	10122101	0	$[[13, 6, 6; 7]]_3$
$[13, 7, 5]_3$	$[13, 3, 9]_3$	1021201	11010212221	0	$[[13, 3, 6; 7]]_3$
$[13, 4, 7]_3$	$[13, 9, 3]_3$	1012010221	10211	0	**$[[13, 9, 3; 4]]_3$**
$[13, 4, 7]_3$	$[13, 6, 6]_3$	1012010221	11110201	0	**$[[13, 6, 3; 4]]_3$**
$[13, 10, 3]_3$	$[13, 4, 7]_3$	1121	1111021001	1	$[[13, 3, 7; 9]]_3$
$[13, 7, 5]_3$	$[13, 7, 5]_3$	1021201	1220221	1	$[[13, 6, 5; 6]]_3$

$[13, 7, 5]_3$	$[13, 4, 7]_3$	1021201	1001201111	1	$[[13, 3, 6; 6]]_3$
$[13, 6, 6]_3$	$[13, 10, 3]_3$	12220021	1211	3	$\mathbf{[[13, 7, 3; 3]]_3}$
$[13, 4, 7]_3$	$[13, 12, 2]_3$	1012010221	11	3	$\mathbf{[[13, 9, 2; 1]]_3}$
$[13, 4, 7]_3$	$[13, 9, 3]_3$	1012010221	12001	3	$\mathbf{[[13, 6, 3; 1]]_3}$

We note that the EAQECCs with boldface parameters in Table 2 have positive net rate which is possible to obtain catalytic codes as shown in [8].

5. Conclusion

Characterization and constructions of linear intersection pairs of negacyclic codes have been presented as well as their applications in constructions of EAQECCs. General characterization and properties of such pairs have been given in terms of their generator polynomials. Explicit constructions of linear s-complementary pairs and linear s-intersection pairs of negacyclic codes have been established for $s \in \{0,1\}$. As applications, constructions of EAQECCs have been given based on these pairs. Some illustrative examples have been presented. It is interesting to study the existence and parameters of linear s-intersection pairs and linear s-complementary pairs of negacyclic codes with arbitrary intersecting dimension as well as the optimality of EAQECCs (see [13]).

Acknowledgments

This research was funded by the National Research Council of Thailand and Silpakorn University under Research Grant N42A650381.

References

[1] Huffman W C and Pless V 2003 *Fundamentals of error-correcting codes* (Cambridge: Cambridge University Press)
[2] MacWilliams F J and Sloane N J A 1977 *The theory of error-correcting codes* (Amesterdam: North-Holland Publishing)
[3] Carlet C, Güneri C, Mesnager S and Özbudak F 2018 Construction of some codes suitable for both side channel and fault injection attacks *Lecture Notes in Comput. Sci.* **11321** 95–107
[4] Carlet C, Güneri C, Mesnager S, Özbudak F, Özkaya B and Solé P 2018 On linear complementary pairs of codes, *IEEE Trans. Inform. Theory* **64** 6583–9
[5] Guenda K, Gulliver T A, Jitman S and Thipworawimon S 2020 Linear *l*-intersection pairs of codes and their applications, *Des. Codes Cryptogr.* **88** 133–52
[6] Hossain M A and Bandi R 2023 Linear *l*-intersection pairs of cyclic and quasi-cyclic codes over a finite field *J. Appl. Math. Comput.* **69** 2901–17
[7] Dougherty S T and Şahinkaya S 2022 On cyclic and negacyclic codes with one-dimensional hulls and their applications *Adv. in Math. of Commun.* 10.3934/amc.2022096
[8] Bosma W, Cannon J and Playoust C 1997 The Magma algebra system. I. The user language *J. Symbolic Comput.* **24** 235–65
[9] Hsich M H, Devetak I and Brun T 2007 General entanglement-assisted quantum error-correcting codes *Phys. Rev.* A **76** 06231
[10] Brun T, Devetak I and Hsieh M H 2014 Catalytic quantum error correction *IEEE Trans. Inform. Theory* **60** 3073–89
[11] Guenda K, Jitman S and Gulliver T A 2018 Constructions of good entanglement-assisted quantum error correcting codes *Des. Codes Cryptogr.* **86** 121–36
[12] Brun T, Devetak I and Hsieh H M 2006 Correcting quantum errors with entanglement *Science* **314** 436–39
[13] Lai C Y and Ashikhmin A 2018 Linear programming bounds for entanglement-assisted quantum error correcting codes by split weight enumerators *IEEE Trans. Inf. Theory.* **64** 622–39.

Effects of strong solar flares on the ionospheric electron density over the Philippines

Zane Nikia C. Domingo and Ernest P. Macalalad*

Department of Physics, Mapúa University, Intramuros, Manila, Philippines

* Corresponding author's email: epmacalalad@mapua.edu.ph

Abstract. Solar flares are capable of generating extreme amounts of ionizing radiation which affects Earth's atmosphere. These are examined most notably in the X-ray and EUV regimes. The eruption of strong solar flares of classes X6.9, X2.25, and M9.7 were recorded, during which the Philippines was on Earth's sunlit side. In this paper, the observation of the effects of solar radiation on the ionospheric electron density is done by quantifying the change in the slant total electron content ($\Delta STEC$). This was calculated by finding the difference between the difference relative STEC and its fitted data. The data were acquired from two GNSS receiver stations (PIMO & PTGG) in the Philippines. Moreover, it was inferred that changes in the electron density are manifest during increased EUV enhancements on the occurrences of the X6.9 and M9.7 flares. The calculated $\Delta STEC$ ranged from $\sim 0.43 - 0.67\ TECU$ during the X6.9 (PIMO) and $\sim 0.45 - 0.71\ TECU$ during the M9.7 (PTGG) flares. This signifies that the contribution of the EUV radiation to the ionospheric response in the E and F ionospheric regions is more immense than that of the X-ray.

1. Introduction

Ionospheric disturbances can occur due to the propulsion of high-energy particles and electromagnetic radiation from solar phenomena. The solar radiation from solar flares in particular is studied at wavelengths of X-ray and enhanced UV (EUV). The absorption of which on Earth's atmosphere gives rise to enhanced ionization in the D, E, and F ionospheric regions. The emitted EUV during flares typically results in the ionosphere and thermosphere being heated up, by which the density at a given altitude and location is increased [1]. The electron density does not only depend on the EUV fluxes but also on the neutral composition and dynamic effects brought by electric fields and neutral winds [2].

Energetic particles precipitating into the ionosphere, and the current within, also increase the density distribution in the upper atmosphere. EUV emission also normally increases for a few hours during strong X-class solar flares, the effect of which on the ionosphere and thermosphere lasts for days [1]. Since EUV, however, varies with time and location on the Sun [3], these may possibly influence the apparent solar emissions that are being recorded. The E and F regions of the upper ionosphere are disturbed due to EUV radiation, while electron density perturbations are created in the D region of the lower ionosphere due to X-rays [4].

Solar flares are classified based on the peak X-ray flux in the 0.1-0.8 nm range. Strong to weak types may include X-, M-, C-, B-, and A-class. It is also worth noting that the development of geomagnetic activities may also be investigated in relation to solar flares and their different intensities. Since these effects are propelled toward Earth's atmosphere, ionospheric responses to geomagnetic storms, if there are any, can be observed.

Journal of Physics: Conference Series **2793** (2024) 012013 doi:10.1088/1742-6596/2793/1/012013

The production of high-energy ionizing radiation during strong solar flares makes it favorable for photoionization in the ionosphere to occur. The layers of ions and free electrons in the ionosphere vary each day and solar cycle (SC). In generating an electron density profile, the electrons in a column of the atmosphere can be summed. The total electron content (TEC) in TEC units (1 TECU = 10^{16} electrons/m^2) is an essential parameter in monitoring the variation of the ionospheric electron density. It is a measure of the number of electrons along a path through which signals are transmitted. TEC can also be mathematically defined as the line integral of the electron density, $N_{e,}$ along the path, l [5].

The thermal energy induced by energetic particles also creates variations in the ionospheric density which affect the path of radio signals. These may also cause positioning errors in navigation data. The Global Navigation Satellite System (GNSS) is currently operating on single or dual frequencies in processing the TEC values. These can be estimated along the radio signal's path between the satellite and ground-based receivers. The time delay in the trans-ionospheric signals can also be obtained by the difference in the signal parameters from the two frequencies [5]. Since satellite and navigation systems highly rely on electron densities, the variability of which that is also coincident with the ionizing radiation due to solar flares can be measured. In order to quantify the changes in the electron density, TEC calculations during the flare occurrences and also the quiet days, that is, the non-flare days are important.

In this paper, the ionospheric effects generated by the strong solar flares on 09 August 2011, 20 April 2022, and 21 April 2022 are analyzed. This involves the strongest X-class in SC 24 and also the (strongest) M-class and (second strongest) X-class solar flares so far in SC 25, during which the Philippines was on the sunlit hemisphere. The objective is to determine the TEC calculations over the Philippine region during the occurrences of solar flares. In addition, the geomagnetic conditions are also shown during the observation of the ionospheric effects.

2. Data and measurements

2.1. Kp, Hpo, Dst indices

The geomagnetic conditions are presented by using three indices that indicate the strength of the geomagnetic disturbances present during the solar flares. The level of activity in the geomagnetic field is shown in a three-hourly observation of the *Kp* index. The data for which were downloaded from https://kp.gfz-potsdam.de/en/data. A relatively new group of indices called the *Hpo* is also utilized. These were accessed from https://kp.gfz-potsdam.de/en/hp30-hp60/data. The derivation of *Hpo* is similar to that of the *Kp*. The half-hourly (30 minutes, *Hp30*) and hourly (60 minutes, *Hp30*) data are indicated by "H", planetary by "p", and open-endedness by "o" [6]. Additional advantages for this include higher temporal resolution and the removal of the upper limit at 9 [6]. The evaluation of indices normally utilizes conversion tables for the geomagnetic disturbance magnitudes and their corresponding integer values. This procedure is done for H likewise for K with the use of a similar frequency distribution [6]. Moreover, in showing the measurement of the strength of the ring current around Earth's geomagnetic field, the Dst values are also used. These would also indicate the emergence of geomagnetic storms. The hourly measurements for which are obtained from NASA/GSFC OMNIWeb Interface, https://omniweb.gsolar flaresc.nasa.gov/form/dx1.html.

2.2. Solar data (X-ray and EUV flux)

Shown in Table 1 are the dates of the solar flares with the time at which these started, maxed (in UT and LT), and ended. Two strong X-class solar flares on SC24/25 and one M-class solar flare on SC25 erupted on 09 August 2011 and 20-21 April 2022.

Table 1. Classification of strong solar flares during the SC24/25 over the Philippines.

Day	Start (UT)	Max (UT)	End (UT)	Max (LT)	Region	Class
09 August 2011	07:48	08:05	08:24	16:05	1263	X6.9
20 April 2022	03:41	03:57	04:04	11:57	2992	X2.25
21 April 2022	01:47	01:59	02:05	09:59	2993	M9.7

Journal of Physics: Conference Series **2793** (2024) 012013 doi:10.1088/1742-6596/2793/1/012013

The Philippines was on the sunlit hemisphere during these events as shown in LT. The short-lived flux variations in the X-ray and EUV wavelengths are also investigated. These solar data were utilized for the observation of the solar flares and were accessed from the Geostationary Operational Environmental Satellite (GOES) archives, https://www.ngdc.noaa.gov/stp/satellite/goes-r.html. Data with 1-minute average fluxes were selected. For the X-ray fluxes, the short (0.5-4Å) and long (1-8Å) wavelengths were included. Also, wavelengths of 121.6 nm and 117.5 nm were arbitrarily chosen for the solar EUV irradiance of the August 2011 and April 2022 solar flares, respectively.

2.3. TEC

There are two GNSS receiver stations that were used, located in Quezon City (PIMO: 14.63571966°N, 121.07773220°E) and Taguig City (PTGG: 14.5354022°N, 121.04126541°E), Philippines. The data for the August 2011 flare were from PIMO and for the April 2022 flares from PTGG. The corresponding RINEX files for which are obtained. The data are extracted from https://cddis.nasa.gov/archive/gnss/data/daily/2022/, which is one of the global data centers of the International GNSS Service (IGS). The L1L2 combination of the carrier phase data was used in the calculations to derive the relative STEC.

The dual-frequency signals transmitted by the GNSS satellites are of frequencies f_1 and f_2 with values $1575.42MHz$ and $1227.60MHz$ [7]. The pseudoranges for which are observed by the GNSS receivers. Equation (1) shows how the relative slant TEC (STEC) along the signal path is determined. Here, ϕ_1 and ϕ_2 are the carrier phase measurements. Their corresponding wavelengths are λ_1 and λ_2 which are equal to $0.190293m$ and $0.24421m$, respectively.

$$Relative\ STEC(TECU) = \frac{1}{40.3\times10^{16}}\frac{f_1^2 f_2^2}{f_1^2-f_2^2}(\lambda_2\phi_2 - \lambda_1\phi_1) \tag{1}$$

The difference relative STEC was then obtained. This was performed by first setting the time interval for which the events are covered. A one-hour interval was set which concentrates on the overall start, maximum, and end of the solar flares. After which, the STEC measurement at the starting time of that interval is subtracted from the entire STEC measurement during that one-hour period.

The elevation angles for the pseudo-random numbers (PRNs) corresponding to the satellite IDs used were also obtained. These were superimposed with the first set of STEC measurements, shown in figure 2. The STEC calculations during the event and quiet days are superimposed as shown in figure 3. Lastly, in quantifying the change in STEC ($\Delta STEC$), a polynomial fit was performed for the calculated STEC. This was based on the STEC within the interval from the start to the end of the solar flare events. Then, the difference between the fitted data and calculated STEC was obtained.

3. Observations and discussions

3.1. Kp, Hpo, Dst indices

The geomagnetic conditions during the X6.9 (9 August), X2.25 (20 April), and M9.7 (21 April) flares are presented in figure 1. Dst measurements are superimposed with the Kp indices and Hp data points. With Kp ≤ 3, the geomagnetic conditions during these days are quiet. There need not be an extension of the days succeeding the solar flares. Hence, with only a 6-hour long observation, the Dst measurements confirmed no emergence of geomagnetic storms. Around the time the X6.9 peaked at 08:05 UT, which in this case is at 8 UT, the Dst value reached $-24nT$. On 20 April, the curve appeared to have reached positive values until its decline at around 4 UT. This was around the time at which the X2.25 peaked. On 21 April, the Dst value peaked at $14nT$ which occurred at 2 UT, near the maximum time of the flare. Overall, measurements of Dst $\geq -50nT$ indicate that the curve hovered around the nominal values and that there are no high geomagnetic activities.

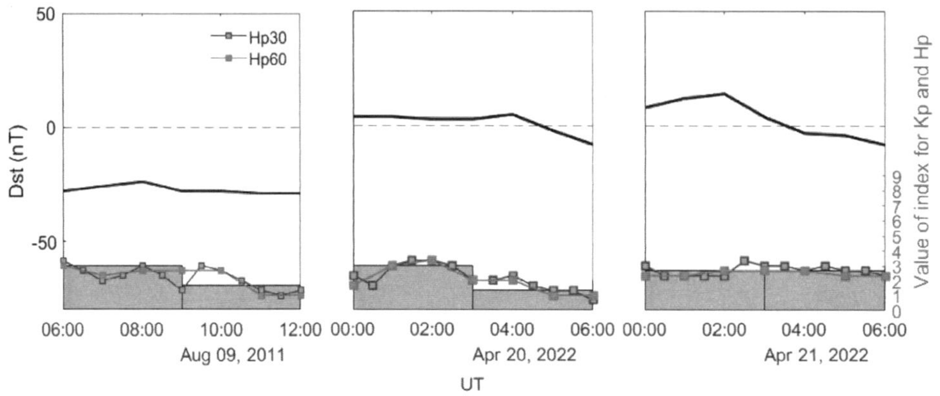

Figure 1. Geomagnetic indices at the interval of 6 hours for each event. The Kp and Hp are superimposed with their combined axis on the right, while the Dst has its axis on the left.

3.2. Solar data (X-ray and EUV flux)

The solar radiation is presented in figure 2, where the first row shows the X-ray fluxes and the second row the EUV fluxes. As validated by the X-ray graphs, the X-class solar flares peaked at $> 10^{-4}\ W/m^2$ and the M-class between 10^{-4} and $10^{-5}\ W/m^2$. The M9.7 also was approaching an X-class. The declines in the X-ray also appeared to be relatively similar among the solar flares, that is, these are gradual. Conversely, the appearance of the peaks was different. This was narrower for the X6.9, followed by the X2.25, then M9.7.

Equal scaling is both applied for the X-ray and EUV flux. The EUV wavelengths, in particular, are scaled by 10^{-3}. EUV Solar irradiances are observed at 121.6 nm for the X6.9 and 117.5 nm for the X2.25 and M9.7. During the X6.9, the EUV enhanced by roughly $0.4340 \times 10^{-3}\ W/m^2$. Whereas only by an estimate of $0.005664 \times 10^{-3}\ W/m^2$ did the EUV flux for the X2.25 increase. The M9.7 had a larger increase compared to the X2.25, in which the estimate was by $0.1145 \times 10^{-3}\ W/m^2$. The apparent difference in the EUV changes between the X2.25 and M9.7 consequently influences the ionospheric changes.

All the EUV enhancements for the solar flares are observed shortly before the X-ray peaks. At around 08:02 UT, the EUV enhanced the largest for the X6.9. This was at 03:54 UT for the X2.25 and 01:55 UT for the M9.7. With these, behaviors in the X-ray and EUV may indeed present differences in terms of the strength at a flare's maximum.

3.3. TEC

In figure 2, the third and remaining rows show the calculated STEC for the available PRNs during those flares. These are superimposed with the elevation angles to confirm the consistency between the position of the satellites and the TEC calculations during the observation time.

It was noted that the negative values would imply the movement of the satellites. When approaching the zenith, the ionosphere through which the signals are transmitted becomes thinner. This in turn affects the amount of TEC being measured. Once the satellites start moving away from the zenith, the path of signal transmission lengthens, resulting in the positive values of the TEC. These behaviors can be explained by the elevation angles of each PRN.

The ionospheric pierce points (IPPs) may also contribute to the measurements. These are the points at which the line of sight between the satellite and receiver intersects with the ionosphere assumed as a thin shell. The farther the IPP trajectories are found from the region where the ionization of the ionosphere takes place more significantly, the less pronounced the effects seen in the TEC calculations might be.

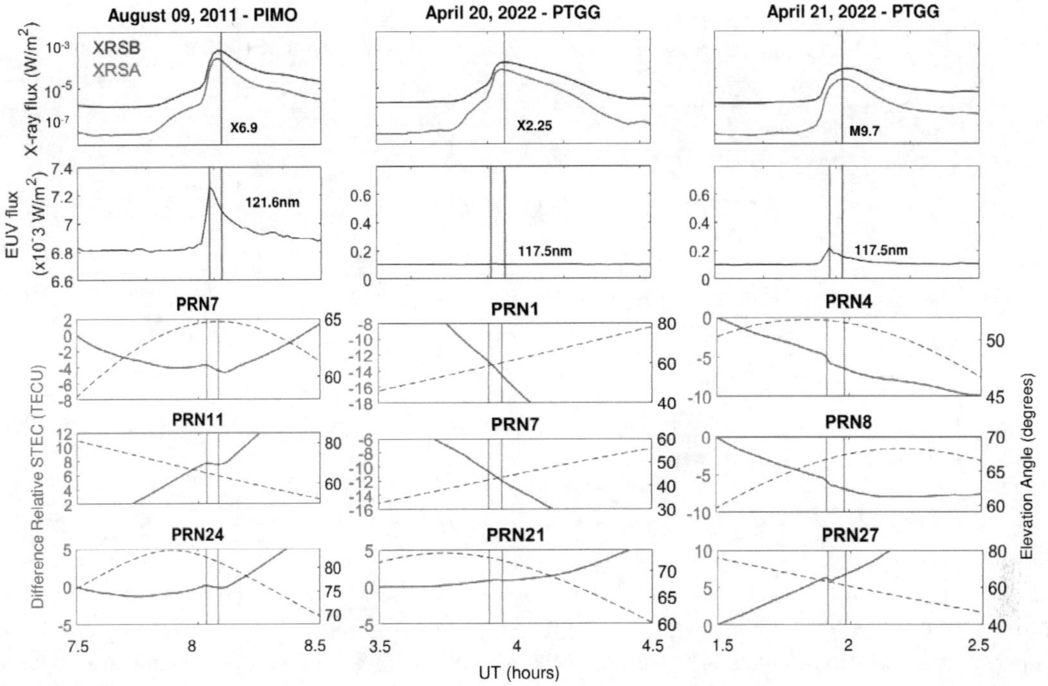

Figure 2. The X-ray flux for the solar flares is presented in the 1st row for the XRSA (0.5-4Å) and XRSB (1-8Å). The 2nd row shows the EUV fluxes for the X6.9 (121.6 nm), X2.25 (117.5 nm), and M9.7 (117.5 nm) with the vertical blue line indicating the time at which the EUV peaked. Included in the 3rd and the succeeding rows are the (red curve) difference relative STEC and the (broken black curve) elevation angles for different PRNs. The vertical black line that goes through each panel from upper to lower indicates the time at which the solar flares reach their maximum.

As seen in figure 2, the EUV for the X6.9 flare peaked at around 08:02 UT (indicated by the blue line), during which the signature in the difference relative STEC is observed. It can be inferred from this that the rapid change in the STEC is due to the enhanced EUV emission in the 121.6 nm wavelength. The STEC response to the EUV enhancement is also evident for the 117.5 nm during the M9.7. There is hardly an EUV increase during the X2.25, which results in no perceptible signature of TEC increase.

Another thing that can be taken into account is the relationship of TEC enhancements with the solar zenith angle (SZA). This is shown in a study made by [8]. At mid and low latitudes, rapid changes in the TEC during solar flares are exhibited in the changing rate of the temporal TEC variations. They demonstrated that for larger SZAs, TEC enhancements are smaller. In addition, the dependence of the TEC enhancements on the SZA at individual IPPs may also be observed.

Furthermore, considering only the X6.9 and M9.7 flares, the measurements in figure 3 include the flare-day and quiet-day STEC. This additionally confirms that the enhancements were present when the flares occurred. The quiet day considered for 9 August occurred on the day before, which was on 8 August. Whereas, the quiet day for 21 April was on 16 April. Again, the signature of the enhancements is evinced in all PRNs as these are clearly distinct from the quiet-day STEC.

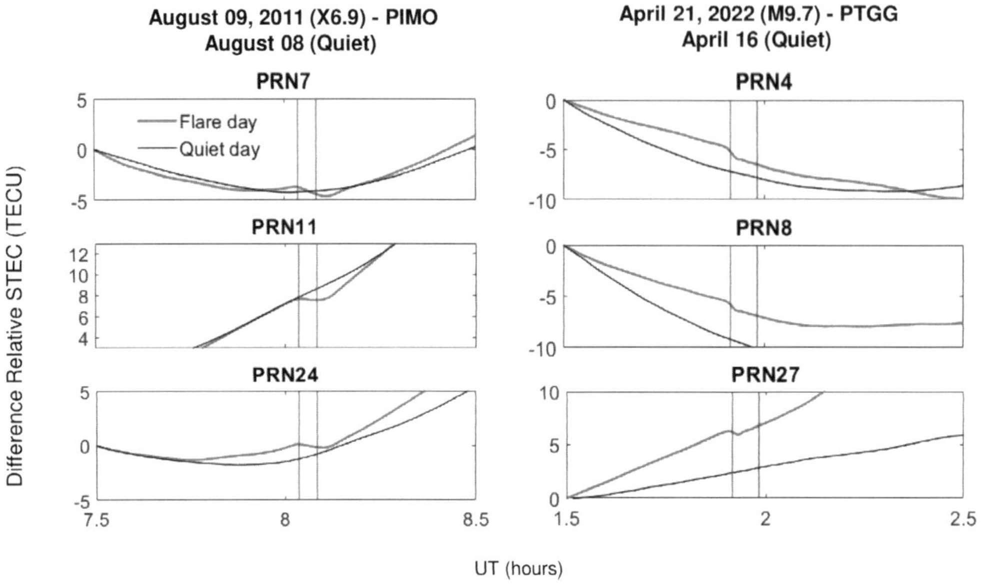

Figure 3. The difference relative STEC for the X6.9 (left column) and M9.7 (right column) are superimposed with quiet days. The vertical blue and black lines indicate EUV and X-ray peaks, respectively.

Aside from considering quiet days to be contrasted with the flare-day STEC, a polynomial fit of the STEC data is also obtained to further quantify $\Delta STEC$. In figure 4, the fitted data for each STEC would show the supposed quiet day during that flare day. For each PRN, there is a corresponding value. During the X6.9 (left panels), PRNs 7,11, and 24 had a change of ~0.62, ~0.67, and ~0.43 TECU, respectively. The maximum values were obtained within the time interval of the start of the solar flares and the EUV peak. When examined closely, these maxima were registered shortly before the EUV peaked. This case is more visible for PRN 11. This may suggest that this corresponded to a different EUV wavelength. Regardless, the rising behavior of the $\Delta STEC$ is evident and is dominantly towards the EUV peak time.

The paper of [9] also investigated the same X6.9 class solar flare observed on 9 August with the response of the equatorial and low-latitude ionosphere. Their calculations showed an enhancement in the vertical TEC (VTEC) of $\sim 2-3\ TECU$ and differential TEC (DVTEC) of $\sim 0.2\, TECU$ in the sunlit hemisphere at the Indian region [9].

The early rise in the $\Delta STEC$ during the X6.9 is also apparent during the M9.7 class flare, as seen in figure 4 (right panels). For PRNs 4,8, and 27, the calculated $\Delta STEC$ are ~0.71, ~0.45, and ~0.50 TECU, respectively. The behavior of the $\Delta STEC$ does not generally present a narrow peak, yet, likewise, it ascends towards the EUV peak.

Moreover, the intensities of solar flares can be measured in either the X-ray or EUV regime. A study by [10] showed high intensity during the 28 October 2003 (X17.2) flare in the EUV and the 04 November 2003 (X28) flare in the X-ray wavelengths, implying a spectral difference between the solar flares. Their results showed that the TEC enhancements were rather lower during the X28 than the X17.2. It was then suggested that this observation was due to the October flare being more intense in the EUV regime, which causes ionization in the E and F ionospheric regions. Since it's the EUV portion of the spectrum that is dominantly responsible for the photoionization in these ionospheric regions, the TEC enhancements would account for this wavelength. Whereas the November flare measured large in the X-ray spectrum, which mostly contributes to the ionization of the D region.

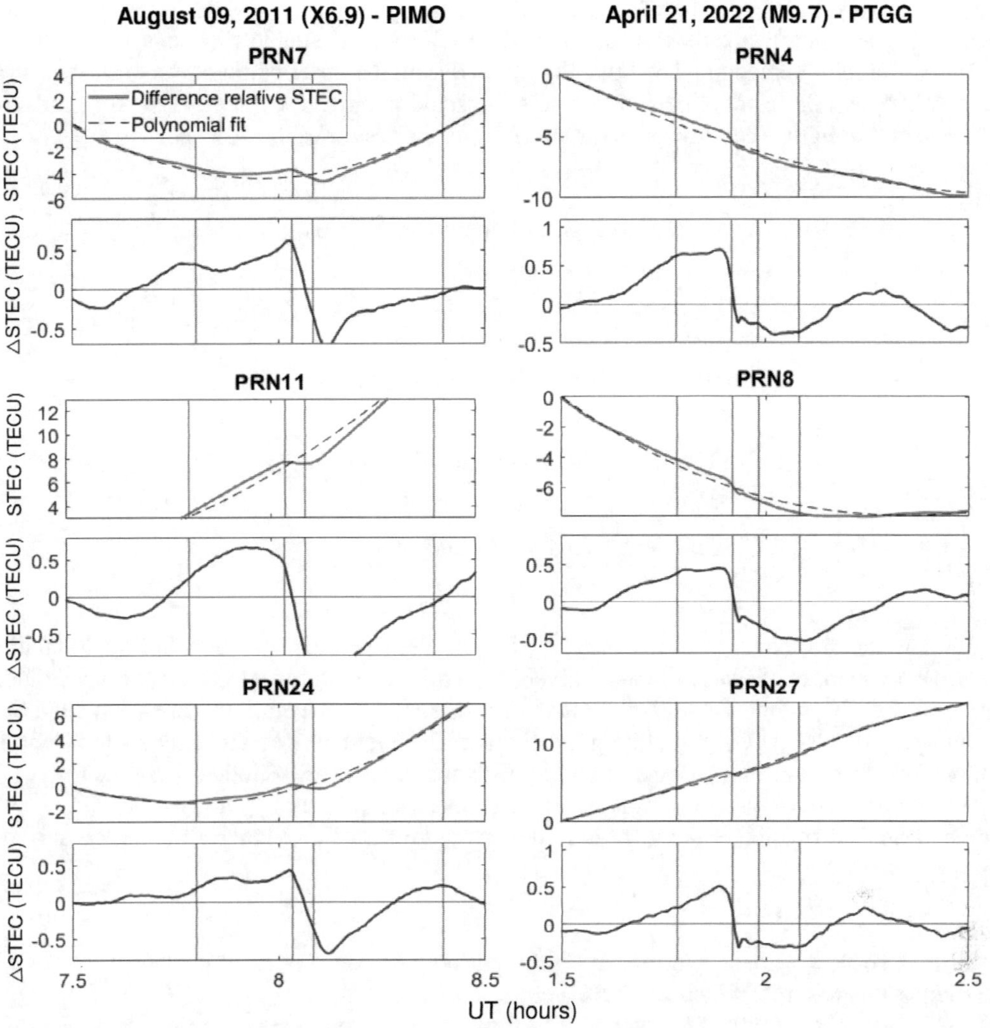

Figure 4. The difference relative STEC (red curve) is superimposed with the polynomial fit (broken black curve) for each PRN during the X6.9 and M9.7 solar flares. The corresponding $\Delta STEC$ (blue curve) between the two calculations is also shown. The vertical blue lines along all panels indicate the EUV peak, while the three vertical black lines indicate the start, peak, and end of the solar flares.

The same observation can be made with the pronounced increases in the STEC observed during the X6.9 and M9.7. Both of which concur with the EUV peaks. Although, there were hardly any significant increases during the X2.25 which corresponded to the EUV peak. The intensity in the X-ray regime is hence not expected to be large.

4. Summary and recommendations

The strongest solar flares on solar cycle 24 and two strong solar flares, so far, on 25 are considered for the observation of the effects on the ionospheric plasma density over the Philippines. An X6.9 flare occurred on 09 August 2011, an X2.25 on 20 April 2022, and M9.7 on 21 April 2022. All events also show no high geomagnetic activities.

Since it is the EUV flux that majorly contributes to the ionization of the ionospheric E and F regions, strong solar flares aren't always expected to produce the largest TEC enhancements. This is reflected in the clear responses of the difference relative STEC to the EUV enhancements found during the X6.9 and M9.7 solar flares. The $\Delta STEC$ calculated for each PRN during the two solar flares also showed an ascent towards the time at which the EUV in the available wavelength peaked.

Furthermore, it is recommended that the SZAs during the observation of the events are included to check other possible factors that would help understand the increases in the TEC for flares that are not of the same strength. The same goes for the IPP trajectories for each PRN.

5. Appendices

Table 2. List of abbreviated words

GNSS	Global Navigation Satellite System
IGS	International GNSS Service
IPP	Ionospheric pierce points
PRN	Pseudo-random number
STEC	Slant total electron content
SZA	Solar zenith angle
TEC	Total electron content

Acknowledgments
The author wishes to express her appreciation for the help provided by Sir Vincent Maglambayan from the Department of Physics, Mapúa University, without which the TEC calculations wouldn't be performed. Also, the author would like to thank the Geostationary Operational Environmental Satellite (https://www.ngdc.noaa.gov/stp/satellite/goes-r.html), GFZ German Research Centre for Geosciences (Helmholtz-Zentrum Potsdam) (https://kp.gfz-potsdam.de/en), NASA/Goddard Space Flight Center OMNIWeb Interface (https://omniweb.gsolar flaresc.nasa.gov/form/dx1.html), and NASA Crustal Dynamics Data Information System (https://cddis.nasa.gov/), for providing all the necessary data in this paper.

References
[1] Lugaz N (2019) Space Weather at Earth and in Our Solar System. In: The Sun as a Guide to Stellar Physics. Elsevier, pp 335–361
[2] Buonsanto MJ (1999) Ionospheric Storms — A Review. Space Sci Rev 88:563–601. https://doi.org/10.1023/A:1005107532631
[3] Thiemann EMB, Eparvier FG, Knoer V, et al (2021) Solar Extreme Ultraviolet Irradiance Uncertainties for Planetary Studies. J Geophys Res Space Phys 126:. https://doi.org/10.1029/2020JA028184
[4] Saharan S, Maurya AK, Dube A, et al (2023) Low latitude ionospheric TEC response to the solar flares during the peak of solar cycle 24. Advances in Space Research 72:3890–3902. https://doi.org/10.1016/J.ASR.2023.07.015
[5] Mendoza MM, Juadines KES, MacAlalad EP, Tung-Yuan H (2019) A Method in Determining Ionospheric Total Electron Content Using GNSS Data for non-IGS Receiver Stations. International Conference on Space Science and Communication, IconSpace 2019-July:186–191. https://doi.org/10.1109/ICONSPACE.2019.8905958
[6] Yamazaki Y, Matzka J, Stolle C, et al (2022) Geomagnetic Activity Index Hpo. Geophys Res Lett 49:. https://doi.org/10.1029/2022GL098860
[7] Gayo WS, Urrutia JD, Temple F, et al (1936) 2-Dimensional Regional Mapping of Ionospheric Total Electron Content Using Kriging Interpolation over the Philippines: Initial Results You may also like Modeling Philippine Stock Exchange Composite Index Using Time Series Analysis 2-Dimensional Regional Mapping of Ionospheric Total Electron Content Using Kriging Interpolation over the Philippines: Initial Results. J Phys Conf Ser 12012. https://doi.org/10.1088/1742-6596/1936/1/012012

[8] Zhang DH, Xiao Z (2003) Study of the ionospheric total electron content response to the great flare on 15 April 2001 using the International GPS Service network for the whole sunlit hemisphere. J Geophys Res Space Phys 108:. https://doi.org/10.1029/2002JA009822

[9] Sripathi S, Balachandran N, Veenadhari B, et al (2013) Response of the equatorial and low-latitude ionosphere to an intense X-class solar flare (X7/2B) as observed on 09 August 2011. J Geophys Res Space Phys 118:2648–2659. https://doi.org/10.1002/JGRA.50267

[10] Tsurutani BT, Judge DL, Guarnieri FL, et al (2005) The October 28, 2003 extreme EUV solar flare and resultant extreme ionospheric effects: Comparison to other Halloween events and the Bastille Day event. Geophys Res Lett 32:. https://doi.org/10.1029/2004GL021475

Journal of Physics: Conference Series **2793** (2024) 012014 doi:10.1088/1742-6596/2793/1/012014

A reliable semi-analytic solution for the Caputo fractional human liver model

Abrar Ansar, Marwan Abukhaled*

Department of Mathematics and Statistics, American University of Sharjah, Sharjah, U.A.E.

Corresponding author's E-mail: mabukhaled@aus.edu

Abstract. This article presents a dependable semi-analytic solution for the Caputo fractional human liver model, employing an enhanced classical differential transform method. Our methodology is direct and produces highly accurate results. We confirm the stability of the equilibrium solution and demonstrate the positivity and boundedness of our solution. The paper includes graphs that juxtapose analytical and numerical simulations for both integer and fractional models, comparing our findings with clinical data and the fourth-order Runge-Kutta method.

1. Introduction

The liver is pivotal in metabolism and detoxification, managing bilirubin from red cell breakdown via the biliary system. Bromosulphthalein (BSP) tests liver function, indicating its ability to process organic anions. Administered intravenously, BSP's hepatic extraction and subsequent excretion into bile reflect liver efficiency. BSP clearance offers insights into liver health, aiding in diagnosing and evaluating treatments. [1, 2].

The field of dynamical systems has evolved, favoring fractional derivatives over integer ones, enhancing model accuracy for systems with memory and long-range interactions. This shift has rejuvenated interest in models initially developed with integer derivatives, now re-examined under the fractional paradigm [3, 4, 5, 6, 7]. Fractional models, superior in depicting complex phenomena like diffusion and viscoelasticity, offer more precise natural process representations [8].

Fractional differential equations pose a challenge in finding exact solutions for various reasons including nonlinearity, the choice of fractional derivative operator, and singularities. Despite this, analytical methods (see [13, 14, 15, 16]), preferred for their precision and error tolerance, alongside adapted classical numerical methods (see [9, 10, 11, 12]) are utilized to find approximate solutions.

In exploring liver function intricacies, Čelechovská utilized mathematical modeling with BSP clinical data obtained to identify key parameters [17]. Calvetti's model further investigates liver's metabolite regulation [18]. Leveraging fractional derivative operator's unique properties, researchers have refined these models, aligning outcomes more closely with clinical data, a testament to the fractional approach's efficacy [19, 20, 21].

This article proposes a semi-analytic Caputo fractional liver model solution via the fractional differential transform method. It covers Caputo derivatives, the fractional liver model, its stability analysis, semi-analytic solution derivation, concluding with results and discussions.

Journal of Physics: Conference Series **2793** (2024) 012014 doi:10.1088/1742-6596/2793/1/012014

2. Preliminaries

This section introduces the fundamental tools of fractional calculus necessary for the remaining parts of this work.

Definition 1. For $m-1<\alpha\leq m$, and $m\in\mathbb{N}$, the αth order Riemann-Liouville fractional integral of the real-valued function $f:\mathbb{R}^+\rightarrow\mathbb{R}$, denoted $J_{t_0}^{\alpha}f(t)$, is defined by

$$J_{t_0}^{\alpha}f(t)=\frac{1}{\Gamma(\alpha)}\int_{t_0}^{t}(t-s)^{\alpha-1}f(s)ds,\ \alpha>0,t>0, \tag{1}$$

and the αth order Caputo fractional derivative of $f(t)$, denoted ${}^{C}D_{t_0}^{\alpha}$ is defined by

$$D_{t_0}^{\alpha}f(t)=J^{m-\alpha}D^{m}f(t). \tag{2}$$

Henceforth, the Caputo fractional derivative and the Riemann-Liouville fractional integral of order α will be denoted D^{α} and J^{α}, respectively.

Lemma 1. *Assume that $f(t)$ is differentiable on $[t_0,b]$ and $m-1<\alpha\leq m$, then*

$$J^{\alpha}D^{\alpha}f(t)=f(t)-\sum_{k=0}^{m-1}\frac{f^{(k)}(t_0)}{k!}(t-t_0)^{k}, \tag{3}$$

and

$$D^{\alpha}J^{\alpha}f(t)=f(t) \tag{4}$$

Definition 2. *Let $F(s)$ be the Laplace transform of $f(t)$, then for $m-1<\alpha\leq m, m\in\mathbb{N}$, the Laplace transform of $D^{\alpha}f(t)$ is given by:*

$$\mathcal{L}\{D^{\alpha}f(t)\}=s^{\alpha}F(s)-\sum_{k=0}^{m-1}s^{\alpha-k-1}f^{(k)}(0) \tag{5}$$

3. Fractional Human Liver Model

In this study, we consider a fraction human liver model, which is a modified version of the model presented by Čelechovská [17], given by

$$\begin{aligned}D^{\alpha}M(t)&=-cM(t)+dN(t),\\ D^{\alpha}N(t)&=cM(t)-(d+f)N(t),\\ D^{\alpha}P(t)&=fN(t),\end{aligned} \tag{6}$$

subject to the following initial conditions: $M(0)=C>0, N(0)=0, P(0)=0$. The variables $M(t)$, $N(t)$, and $P(t)$ represent the amounts of bromsulphthalein (BSP) in the blood, liver, and bile at time t. In addition, c represents the transfer rate of BSP from the blood to the liver, f represents the transfer rate of BSP from the liver to the bile, and d represents the reflux rate of BSP from the liver into the blood.

By summing the equations in (6) and using the properties of the Caputo fractional derivative, it follows that

$$D_{t}^{\alpha}\left(M(t)+N(t)+P(t)\right)=0, \tag{7}$$

and hence $t\geq 0$, we have

$$M(t)+N(t)+P(t)=C, \tag{8}$$

Journal of Physics: Conference Series 2793 (2024) 012014 doi:10.1088/1742-6596/2793/1/012014

This signifies the conservation of the total amount of bromsulphthalein (BSP) in the system. That is, the total amount of BSP in the blood, liver, and bile remains constant over time. Since

$$P(t) = C - N(t) - M(t), \tag{9}$$

we can focus on the simplified system:

$$\begin{aligned} D^{\alpha}M(t) &= -cM(t) + dN(t), \\ D^{\alpha}N(t) &= cM(t) - (d+f)N(t), \end{aligned} \tag{10}$$

with the initial conditions:

$$M(0) = C > 0, \quad N(0) = 0. \tag{11}$$

4. Mathematical analysis of the model

Since c, d, and f are non-negative real numbers, then the only equilibrium solution of the homogeneous counterpart of the fractional model given in (10) is $\mathbf{0} = (0, 0)$.

Theorem 1. *The equilibrium solution* $\mathbf{0}$ *of system* (10) *is asymptotically stable if*

$$(c + d + f)^2 \geq 4(cf). \tag{12}$$

Proof. It is straightforward to see that the eigenvalues of the Jacobian matrix of (10) are given by

$$\lambda_i = \frac{-A + (-1)^{i+1}\sqrt{A^2 - 4(cf)}}{2}, \tag{13}$$

where $i = 1, 2$, and $A = c + d + f$. The system (10) is, then, asymptotically stable if λ_1 and λ_2 are negative. Therefore $A^2 \geq 4cf$.

Theorem 2. *The region* $\Lambda^+ = \{(M(t), N(t), P(t)) : M > 0, t, N, P \geq 0\}$ *is a positive invariant for the fractional system* (6).

Proof. Notice that

$$\begin{aligned} D_t^{\alpha}M(t)\Big|_{M=0} &= dN(t) \geq 0, \\ D_t^{\alpha}N(t)\Big|_{N=0} &= cM(t) \geq 0, \\ D_t^{\alpha}P(t)\Big|_{P=0} &= fN(t) \geq 0, \end{aligned} \tag{14}$$

therefore, by the fractional mean value theorem [23], Λ^+ is a positive invariant.

To ensure that the solution of the system (6) is bounded, we need to show that the positive invariant property is preserved under the constraint that $M + N + P = C$.

Theorem 3. *The region* $\Lambda = \{(M, N, P) : M > 0, t, N, P \geq 0, M + N + P = C\}$ *is a positive invariant set for the system* (6).

Proof. For the linear fractional model (10), the explicit solution, in terms of Mittag-Leffler function, is given by [24]:

$$M(t) = \frac{(AC)(E_{\alpha}(\lambda_2 t^{\alpha}) - E_{\alpha}(\lambda_1 t^{\alpha}))}{2\sqrt{D}} + \frac{C}{2}(E_{\alpha}(\lambda_2 t^{\alpha}) - E_{\alpha}(\lambda_1 t^{\alpha})) > 0, \tag{15}$$

$$N(t) = \frac{-cC(E_{\alpha}(\lambda_2 t^{\alpha}) - E_{\alpha}(\lambda_1 t^{\alpha}))}{\sqrt{D}} t^{1-\alpha} \geq 0, \tag{16}$$

Journal of Physics: Conference Series **2793** (2024) 012014 doi:10.1088/1742-6596/2793/1/012014

where $A = c+d+f$, and λ_1, λ_2 are the eigenvalues of the Jacobian given in (13), $D = A^2 - 4cf$, and $E_\alpha(t)$ is the Mittag-Leffler function given by

$$E_\alpha(t) = \sum_{k=0}^{\infty} \frac{t^k}{\Gamma(\alpha k+1)},\ t, \alpha \in \mathbb{C}\,; \Re(a) > 0 \tag{17}$$

We just need to show that $P(t) \geq 0$ for $t > 0$. By applying the Laplace transform to the third equation in the system (6), we have:

$$\begin{aligned} \mathcal{L}(D_t^\alpha P) &= \mathcal{L}(fN), \\ s^\alpha \bar{P}(s) - s^{\alpha-1}P(0) &= f\bar{N}(s), \end{aligned} \tag{18}$$

where $\mathcal{L}(P(t)) = \bar{P}(s)$ and $\mathcal{L}(N(t)) = \bar{N}(s)$ Therefore

$$\bar{P}(s) = f\frac{1}{s^\alpha}\bar{N}(s). \tag{19}$$

Taking inverse Laplace and employing the convolution theorem leads to

$$P(t) = \frac{f}{\Gamma(\alpha)} \int_0^t (t-\tau)^{\alpha-1} N(\tau)\, d\tau \geq 0. \tag{20}$$

Hence, the solution region Λ is positively invariant, and the solution is bounded.

The existence and uniqueness of the solution to the linear human liver model (10) can be established based on Theorem 3.1 in [24, 25].

5. Fractional differential transform method (FDTM)

Since 1986, the differential transform method (DTM) has been intensively used to produce semi-analytic solutions to ordinary and partial differential equations [26]. The fractional version of the DTM (FDTM) was first proposed by Arikoglu and Ozkol in 2007 [27]. Since then, the FDTM and its modifications have been successfully employed by many researchers for solving fractional integral and differential systems (see [28, 29] and the references therein).

Let $f(t)$ be a continuous analytic function, and the fractional power series expansion is expressed as:

$$f(t) = \sum_{k=0}^{\infty} F(k)\,(t-t_0)^{k/\alpha}\,, \tag{21}$$

where $F(k)$ is the fractional differential transformation of $f(t)$ defined by:

$$F(k) = \begin{cases} \frac{1}{(k/\alpha)!} D_t^{k/\alpha}\Big|_{t=t_0}, & \text{if } k/\alpha \in \mathbb{Z}^+,\ k = 0, 1, \ldots, (q\alpha - 1) \\ 0, & \text{if } k/\alpha \notin \mathbb{Z}^+ \end{cases}. \tag{22}$$

The following form of the Caputo fractional derivative for a fractional-order q follows immediately from Definition 2 and lemma 1:

$$D_{t_0}^q f(t) = \frac{1}{\Gamma(m-q)} D^m \left\{ \int_{t_0}^t \left(f(t) - \sum_{k=0}^{m-1} \frac{(t-t_0)^k}{k!} f^{(k)}(t_0) \right) (t-x)^{m-q-1} \mathrm{d}t \right\}. \tag{23}$$

Some of the most important properties of the FDTM are given in the following theorems [27]:

Theorem 4. *Assume that $f(x)$ and $g_i(x)$, $i = 1, 2, \cdots, n$, are continuous analytic functions, then*

(i) If $f(x) = g_1(x) \pm g_2(x) \pm \cdots \pm g_n(x)$, *then* $F(k) = G_1(k) \pm G_2(k) \pm \cdots \pm G_n(k)$.

(ii) If $f(x) = \prod_{i=1}^{n} g_i(x)$, *then*

$$F(k) = \sum_{k_{n-1}=0}^{k} \cdots \sum_{k_2=0}^{k_3} \sum_{k_1=0}^{k_2} G_1(k_1) G_2(k_2 - k_1) \ldots G_n(k - k_{n-1}).$$

Theorem 5. If $f(x) = (x - x_0)^p$, then $F(r) = \delta(r - \alpha p)$, where

$$\delta(r) = \begin{cases} 1 & \text{if } r = 0 \\ 0 & \text{if } r \neq 0 \end{cases}.$$

Theorem 6. If $f(x) = D^q_{x_0} g(x)$, then

$$F(r) = \frac{\Gamma(q + 1 + r/\alpha)}{T(1 + r/\alpha)} G(r + \alpha q).$$

Now, let's consider the fractional human liver model in (10). From Equations (21) and (22), the truncated series solution for system (10), derived by the FDTM has the form

$$\begin{aligned} M[t] &= \sum_{j=0}^{k} M[j] t^{j\alpha}, \\ N[t] &= \sum_{j=0}^{k} N[j] t^{j\alpha}, \end{aligned} \tag{24}$$

where the coefficients $M[j]$ and $N[j]$, for $0 \leq j \leq k$, are given by:

$$\begin{aligned} M[j+1] &= \frac{\Gamma(\alpha j + 1)}{\Gamma(\alpha j + \alpha + 1)} (-cM[j] + dN[j]), \\ N[j+1] &= \frac{\Gamma(\alpha j + 1)}{\Gamma(\alpha j + \alpha + 1)} (cM[j] - (d + f)N[j]). \end{aligned} \tag{25}$$

6. Results and discussion

At first, we utilized the fractional differential transformation method to solve system (10) subject to the initial conditions $M_0 = 250$ and $N_0 = 0$ for the integer case $\alpha = 1$. In comparison with the fourth-order Runge-Kutta method (RK4), we found that the maximum absolute differences between FDTM and RK4 for $M(t)$ and $N(t)$ are approximately 2.69×10^{-5} and 1.31×10^{-5}, respectively. Figure 1 depicts the numerical RK4 and the analytical FDTM curves for both $M(t)$ and $N(t)$ from which we infer that the proposed semi-analytic method is highly accurate and reliable in offering an insight into the liver extraction and excretion of BSP from blood into bile.

Figure 2 visually presents various fractional curves of $M(t)$ and $N(t)$ for various values of α. These fractional curves exhibit behavior consistent with the integer curve ($\alpha = 1$), providing visual confirmation of the hereditary property associated with the fractional derivative operator, as discussed in the introduction. Biologically, interpreting the consistency of behavior across varying values of α with the integer curve necessitates an understanding of the liver's regenerative

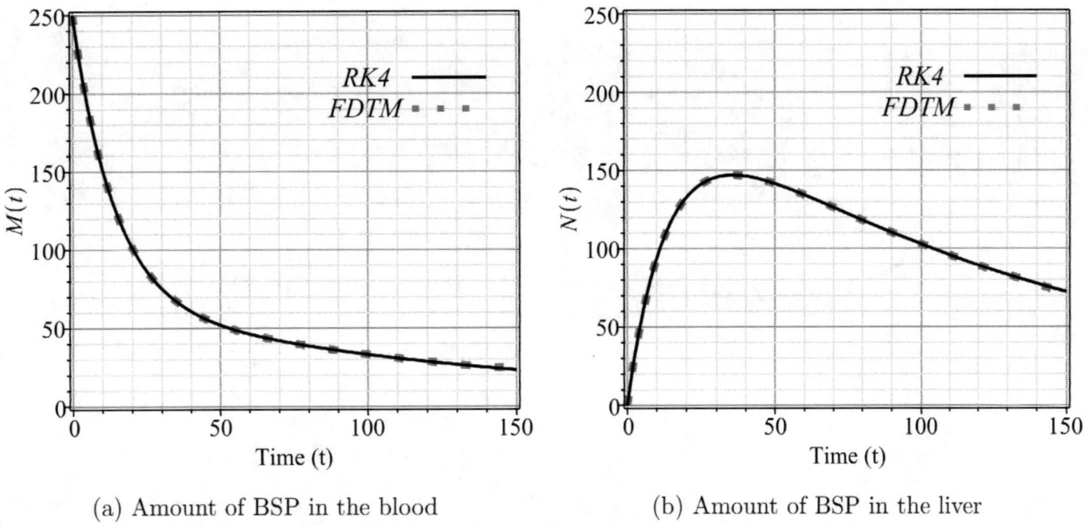

Figure 1: FDTM and RK4 solution curves of the system (10) for the integer case $\alpha = 1$.

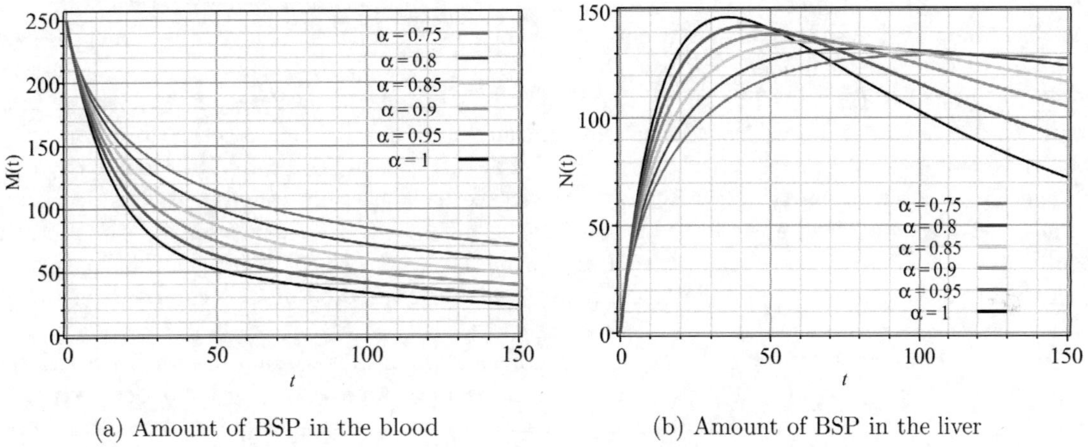

Figure 2: DTM curves for the two groups of the Human liver model for different fractional values of α.

capabilities and the progression mechanisms of liver diseases. For instance, as α approaches 1, our model mimics the classical, memoryless behavior of liver function and disease progression, which could be likened to acute conditions where past states have minimal impact on current dynamics.

Previous studies have shown that the fractional curve with $\alpha \approx 0.98$ reveals a more accurate corresponds with the clinical data than the classical integer-order derivative model [19, 24]. Our study confirm this conclusion as portrayed in Figure 3.

7. Conclusion

In this article, we explore a fractional human liver model employing the Caputo fractional derivative. The model serves to detect liver diseases based on a clinical test, measuring the remaining of the bromsulphthalein in the blood after injection.Validation of the model

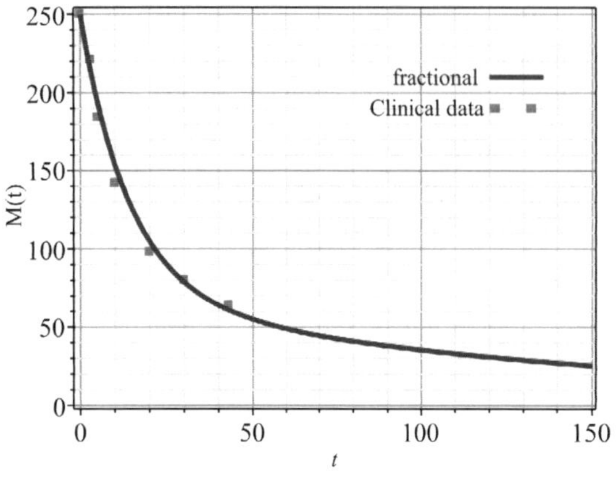

(a) Amount of BSP in the blood

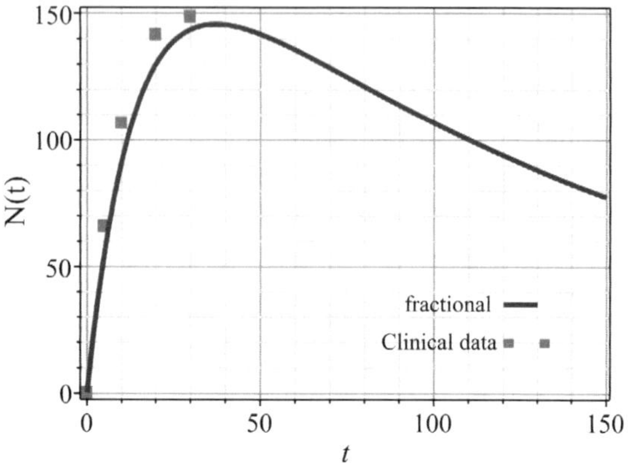

(b) Amount of BSP in the liver

Figure 3: DTM curves for the two groups of the Human liver model for different fractional values of α.

was achieved using clinical data. The utilization of fractional derivative operators, known for describing systems with memory and long-range dependencies, enhances its effectiveness compared to its integer model counterpart.

We established the stability of the equilibrium solution, as well as the positivity, and boundedness of the solution. Leveraging the efficient fractional differential transform approach, we derived a series solution that demonstrated high accuracy when compared to the fourth-order Runge-Kutta solution, particularly for the integer case ($\alpha = 1$). Solutions for various fractional values of α were obtained and plotted alongside the integer solution curve on the same graph, confirming the preservation of the hereditary property, which plays a crucial rule in various biological systems, where the prediction of future states of fractional models can be driven from present states.

In conclusion, our study not only provides a robust mathematical framework for the fractional human liver model but also showcases the advantages of employing fractional derivatives in modeling systems with intricate dynamics. The accuracy and versatility of our proposed simple approach open avenues for extending its applications in understanding and predicting complex biological phenomena.

References

[1] Abdel-Misih SR and Bloomston M 2010 Liver anatomy *Surg Clin North Am* **90**(4) pp 643–53. doi: 10.1016/j.suc.2010.04.017
[2] Gowda S, Desai PB, Hull VV, Math AA, Vernekar SN and Kulkarni S 2009 A review on laboratory liver function tests *Pan Afr Med J* **3** 17
[3] Ullah S, Khan M, Farooq M, Hammouch Z and Baleanu D 2020 A fractional model for the dynamics of tuberculosis infection using Caputo-Fabrizio derivative *Discrete Cont Dyn-S* **13**(3) pp 975–993
[4] Abukhaled M and Khuri S 2023 RLC electric circuit model of fractional order: A Green's function approach *International Journal of Computer Mathematics* DOI: 10.1080/00207160.2023.2203787
[5] Ansar A, Abukhaled M and Khuri S 2023 A Multi-step Differential Transform Approach for a Nonlinear Fractional COVID–19 Pandemic Model *Proc. Int. Conf. on Fractional Differentiation and Its Applications (ICFDA)* IEEE pp 1-5
[6] Cevikel A, Bekir A, Abu Arqub O and Abukhaled M 2022 Solitary wave solutions of Fitzhugh–Nagumo-type equations with conformable derivatives *Frontiers in Physics* **10** 1028668
[7] Rabah F, Abukhaled M and Khuri S 2022 Solution of a complex nonlinear fractional biochemical reaction model *Mathematical and Computational Applications* **27**(3) 45

Journal of Physics: Conference Series **2793** (2024) 012014 doi:10.1088/1742-6596/2793/1/012014

[8] Giona M, Cerbelli S and Roman H 1992 Fractional diffusion equation and relaxation in complex viscoelastic materials *Physica A: Statistical Mechanics and its Applications* **191** pp 449–453
[9] Du Q, Gunzburger M and Ju S 2005 Conforming and nonconforming finite element methods for solving the Riesz fractional elliptic equation *SIAM Journal on Numerical Analysis* **47**(4) pp 2388–2410
[10] Rehman M and Khan R 2011 The Legendre wavelet method for solving fractional differential equations *Communications in Nonlinear Science and Numerical Simulation* **16**(11) pp 4163–4173
[11] Arshad M, Baleanu D, Riaz M and Abbas M 2020 A novel 2-stage fractional Runge–kutta method for a time-fractional logistic growth model *Discrete Dynamics in Nature and Society* **2020** pp 1–8
[12] Shoaib M, Abukhaled M, Kainat S, Nisar K, Raja, M and Zubair G 2022 Integrated Neuro-Evolution-Based Computing Paradigm to Study the COVID-19 Transposition and Severity in Romania and Pakistan *Int J Comput Intell Syst* **15** 80
[13] Abu Arqub O, Tayebi S, Momani S and Abukhaled M 2023 Adaptation of the Novel Cubic B-Spline Algorithm for Dealing with Conformable Systems of Differential Boundary Value Problems concerning Two Points and Two Fractional Parameters *Journal of Function Spaces* **2023** Article ID 5322092
[14] Danane J, Allali K and Hammouch Z 2020 Mathematical analysis of a fractional differential model of HBV infection with antibody immune response *Chaos, Solitons & Fractals* **136** 109787
[15] Abu Arqub O and Maayah B 2019 Modulation of reproducing kernel Hilbert space method for numerical solutions of Riccati and Bernoulli equations in the Atangana-Baleanu fractional sense *Chaos, Solitons & Fractals* **125** pp 163–170
[16] Veeresha P, Prakasha D and Baskonus H 2019 Solving smoking epidemic model of fractional order using a modified homotopy analysis transform method *Mathematical Sciences* **13** pp 115–128
[17] Čelechovská L 2004 A simple mathematical model of the human liver *Appl Math-Czech* **49**(3) pp 227–246
[18] Calvetti D, Kuceyeski A and Somersalo E 2008 A mathematical model of liver metabolism: from steady state to dynamic, J Phys: Conf Ser **124**(1) p. 012012
[19] Baleanu D, Jajarmi A, Mohammadi H and Rezapour S 2020 A new study on the mathematical modelling of human liver with Caputo–Fabrizio fractional derivative *Chaos, Solitons & Fractals* **134** P. 109705
[20] Azeem M, Farman M, Abukhaled M, Nisar K S and Akgul A 2023 Epidemiological analysis of human liver model with fractional operator *Fractals* **31**(4) p. 2340047
[21] Rashid S, Jarad F and Ahmad A 2022 A novel fractal-fractional order model for the understanding of an oscillatory and complex behavior of human liver with non-singular kernel *Results in Physics* **35** p. 105292
[22] Kilbas A, Srivastava H and Trujillo J 2006 Theory and applications of fractional differential equations (Amsterdam: Elsevier)
[23] Odibat Z and Shawagfeh N 2007 Generalized Taylor's formula *Applied Mathematics and Computation* **186** pp 286—293
[24] Ameen I G, Sweilam N H and Ali H M 2021 Fractional-order model of human liver: Analytic-approximate and numerical solu-tions comparing with clinical data *Alexandria Engineering Journal* **60** pp 4797–4808
[25] Lin W 2007 Global existence theory and chaos control of fractional differential equations *Journal of Mathematical Analysis and Applications* **332** pp 709–726
[26] Zhou J 1986 Differential transformation and its applications for electrical circuits (Wuhan: Huarjung University Press)
[27] Arikoglu A and Ozkol I 2007 Solution of fractional differential equations by using differential transform method *Chaos, Solitons & Fractals* **34** pp 1473–1481
[28] Ertürk V S and Momani S 2008 Solving systems of fractional differential equations using differential transform method *J. Comput. Appl. Math* **215** pp 142–151
[29] Abukhaled M, Khuri S and Rabah F 2033 Solution of a nonlinear fractional COVID-19 model *Int. J. Numerical Methods for Heat & Fluid Flow* **32** pp 3657–3670

Journal of Physics: Conference Series **2793** (2024) 012015 doi:10.1088/1742-6596/2793/1/012015

Dynamical concurrence of initially separable bipartite system immersed in N-qubit environment

R Alporha*, **L J Sese**, **R Gammag**

Department of Physics, Mapúa University, Intramuros, Manila, Philippines

E-mail: renzoalporha7@gmail.com*

Abstract. In this paper, we calculate the concurrence for two entangled qubits that are immersed in either a homogeneous or a heterogeneous N-qubit environment. The concurrence for the discrete homogeneous environment was found to behave periodically depending on the interaction strengths $\omega_{\mathcal{S}_1\mathcal{S}_2}$ and Γ. For the case of a heterogeneous environment, we set the interaction strength $\Gamma_{\mathcal{S}_i\mathcal{E}_k}$ as uniformly random. This case mimics of a noisy environment as it illustrates an asymptotic decay of entanglement. The sudden death behavior found in the heterogenous case becomes more prominent for larger N number of qubit environments, and no other entanglement is found as time $t \to \infty$.

1. Introduction

Entanglement is a quantum property that can be formed from the interaction of two systems, where the resulting wave function is inseparable and cannot be factored as a description of its local constituents [1]. Due to its close connection with quantum information theory, entanglement finds application in quantum key distribution, quantum dense coding, quantum teleportation, and entanglement swapping [2].

Unfortunately, entanglement is highly sensitive to interactions with the environment. It is often exposed to a noisy environment, causing the entanglement to undergo either asymptotic decay or sudden death [2, 3]. This decay is caused by another quantum phenomenon called decoherence, where the environment forces the entangled quantum system to decohere into a classical system. However, some studies suggest that environment can also maximize the system entanglement [4, 5]. Using entanglement, computation speed and exchange of information between qubits will exponentially speed up compared to the classical bits due to the superposition principle. As stated earlier, decoherence causes the quantum system to evolve classically. This results in a slower information exchange [2]. The rate of decoherence depends on the number of environments interacting with the system [6]. In fact, many researchers are shifting from entanglement to other approaches such as quantum discord, due to its cheaper and easier lab maintenance [7].

The study of Sese et al. suggests that for a qubit immersed in a homogenous N-qubit environment, there is an exact recurrence for the qubit system to recover its dissipated initial information from the environment. Moreover, the recurrence time was found to be dependent on the parity of N [8]. On the other hand, many recent studies showed that the concurrence dynamics for a bipartite system will always undergo asymptotic decay in a dissipative environment [9, 10]. Most literatures [9–12] often consider a continuous environment for bipartite entanglement. However, in this study, we illustrate a discrete case of an N-qubit homogeneous and heterogeneous environment in the system of an initially separable bipartite system using concurrence as one of the standard measurements for bipartite entanglement. From reference [8], we hypothesize a similar outcome, but in this case, an exact revival in separability and periodicity for a bipartite system in a homogeneous environment. Similar to references [9–12], we also expect an asymptotic decay in discrete heterogeneous environments. We use of the Markovian condition for both environments in acquiring measurements where the present measurement is independent of past measurements.

Following the gaps that we have presented, this paper is arranged as follows: in Section 2, the interaction potential used is introduced. This is followed by a derivation of the density matrix of the System of Interest (SOI) from the composite system. In Section 3, we make use of the acquired SOI density matrix to calculate and analyze the dynamics of the system entanglement using concurrence in both homogeneous and heterogeneous environments. And lastly, we encapsulate things we have discussed in Section 4.

2. Interaction structure

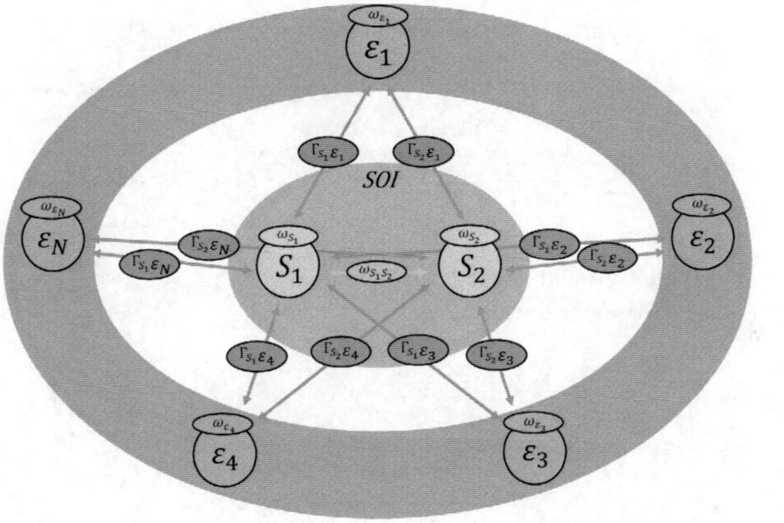

Figure 1: The model for central two-qubit system (SOI) interacting with N number of independent qubit environment.

We will utilize a composite system composed of two interacting systems, namely the SOI and a non-interacting N-qubit environment $\mathcal{E}_k$. The entanglement is set between the two initially separable qubits in SOI. The SOI is submerged in N-qubit environment and the potential can be expressed by placing an interaction Hamiltonian $\hat{H}_{\text{int}}$ in equation (1).

The Hamiltonian of the composite system is expressed as

$$\hat{H} = \hat{H}_{\mathcal{S}} + \hat{H}_{\mathcal{E}} + \hat{H}_{\text{int}}. \tag{1}$$

This model assumes that the three terms are diagonal and $\hat{H}_{\text{int}}$ commutes with both $\hat{H}_{\mathcal{S}}$ and $\hat{H}_{\mathcal{E}}$, that is, $[\hat{H}_{\mathcal{S}}, \hat{H}_{\text{int}}] = 0$ and $[\hat{H}_{\mathcal{E}}, \hat{H}_{\text{int}}] = 0$. Here the Hamiltonian of the SOI $\hat{H}_{\mathcal{S}}$, N-qubit

Journal of Physics: Conference Series 2793 (2024) 012015 doi:10.1088/1742-6596/2793/1/012015

environment $\hat{H}_{\mathcal{E}}$, and system-environment interaction H_{int} are defined as

$$\begin{aligned}\hat{H}_{\mathcal{S}} &= \frac{\hbar}{2}\omega_{\mathcal{S}_1}\hat{\sigma}^z_{\mathcal{S}_1} + \frac{\hbar}{2}\omega_{\mathcal{S}_2}\hat{\sigma}^z_{\mathcal{S}_2},\\ \hat{H}_{\mathcal{E}} &= \frac{\hbar}{2}\sum_{k=1}^{N}\omega_{\mathcal{E}_k}\hat{\sigma}^z_{\mathcal{E}_k},\\ \hat{H}_{\text{int}} &= \frac{\hbar}{2}\omega_{\mathcal{S}_1\mathcal{S}_2}\hat{\sigma}^z_{\mathcal{S}_1}\hat{\sigma}^z_{\mathcal{S}_2} + \frac{\hbar}{2}\sum_{k=1}^{N}\Gamma_{\mathcal{S}_1\mathcal{E}_k}\hat{\sigma}^z_{\mathcal{S}_1}\hat{\sigma}^z_{\mathcal{E}_k} + \frac{\hbar}{2}\sum_{k=1}^{N}\Gamma_{\mathcal{S}_2\mathcal{E}_k}\hat{\sigma}^z_{\mathcal{S}_2}\hat{\sigma}^z_{\mathcal{E}_k}.\end{aligned} \tag{2}$$

The coefficients $\omega_{\mathcal{S}_i}$, $\omega_{\mathcal{S}_1\mathcal{S}_2}$, $\omega_{\mathcal{E}_k}$, and $\Gamma_{\mathcal{S}_i\mathcal{E}_k}$ are the coupling strengths and the $\hat{\sigma}^z_{\mathcal{S}_k}$ and $\hat{\sigma}^z_{\mathcal{E}_k}$ are the Pauli-spin operator in the z-basis of the corresponding k^{th} system or environment, respectively. These are aligned with the study of Cucchietti et al. [6] but without the off-diagonal self-Hamiltonian. From this point onwards, we will $\hbar = 1$. Moreover, we separately define the initial state of the system $|\psi_{\mathcal{S}_k}(0)\rangle$ and the environment $|\psi_{\mathcal{E}_j}(0)\rangle$ with its coefficients and orthonormal bases $\{|0\rangle, |1\rangle\}$ as

$$|\psi_{\mathcal{S}_k}(0)\rangle = a_0|0\rangle_k + a_1|1\rangle_k \quad ; \quad |\psi_{\mathcal{E}_j}(0)\rangle = \alpha_0|0\rangle_j + \beta_1|1\rangle_j. \tag{3}$$

Hence, the density matrix of the composite system will take the form

$$\hat{\rho}(t) = \sum_{\substack{l,\,m,\\ n,\,o}=0}^{1} a_{lm}a^*_{no}\nu_{lmno}|lm\rangle\langle no|\bigotimes_{k=1}^{N}|\psi_{\mathcal{E}k,lm}(t)\rangle\langle\psi_{\mathcal{E}k,no}(t)|. \tag{4}$$

Since we already established that the entanglement is only in SOI, we have to trace out the degrees of freedom of the qubit environment,

$$\hat{\rho}_{\mathcal{S}}(t) = \sum_{\substack{k,\,l,\\ m,\,n}=0}^{1} a_{kl}a^*_{mn}r_{klmn}(t)|kl\rangle\langle mn|, \tag{5}$$

where the expressions of $r_{klmn}(t)$ are to be determined in Section 3.

Now we can make use of the explicit form of concurrence in our SOI reduced density matrix $\hat{\rho}_{\mathcal{S}}(t)$ given by Wooters [13],

$$C(\hat{\rho}_{\mathcal{S}}) = \max\{0, \lambda_1 - \lambda_2 - \lambda_3 - \lambda_4\}. \tag{6}$$

Here, λ_i are the square roots of the eigenvalues of matrix $R = \rho_{\mathcal{S}}\tilde{\rho_{\mathcal{S}}}$ arranged in decreasing order, and $\tilde{\rho_{\mathcal{S}}}$ is a spin flipped operator expressed as $\tilde{\rho_{\mathcal{S}}} = (\sigma^y \otimes \sigma^y)\rho^*_{\mathcal{S}}(\sigma^y \otimes \sigma^y)$.

3. Results and discussion

In both homogeneous and heterogeneous cases, we obtained that the diagonal terms of $r_{klmn}(t)$ are unity. The off-diagonal elements of $r_{klmn}(t)$ are almost periodic functions [14] and are the

Journal of Physics: Conference Series **2793** (2024) 012015 doi:10.1088/1742-6596/2793/1/012015

decoherence factors of reduced density matrix $\hat{\rho}_{\mathcal{S}}(t)$. The decoherence factors are:

$$r_{\substack{0010\\0111}} = r^*_{\substack{0100\\1101}} = \exp[-it(\omega_{\mathcal{S}_1} \pm \omega_{\mathcal{S}_1\mathcal{S}_2})] \prod_{k=1}^{N} (|\alpha_k|^2 \exp[-it\Gamma_{\mathcal{S}_1\mathcal{E}_k}] + |\beta_k|^2 \exp[it\Gamma_{\mathcal{S}_1\mathcal{E}_k}]),$$
$$r_{\substack{0001\\1011}} = r^*_{\substack{0100\\1110}} = \exp[-it(\omega_{\mathcal{S}_2} \pm \omega_{\mathcal{S}_1\mathcal{S}_2})] \prod_{k=1}^{N} (|\alpha_k|^2 \exp[-it\Gamma_{\mathcal{S}_2\mathcal{E}_k}] + |\beta_k|^2 \exp[it\Gamma_{\mathcal{S}_2\mathcal{E}_k}]),$$
$$r_{\substack{0011\\0110}} = r^*_{\substack{1100\\1001}} = \exp[-it(\omega_{\mathcal{S}_1} \pm \omega_{\mathcal{S}_2})] \prod_{k=1}^{N} (|\alpha_k|^2 \exp[-it(\Gamma_{\mathcal{S}_1\mathcal{E}_k} \pm \Gamma_{\mathcal{S}_2\mathcal{E}_k})] + |\beta_k|^2 \exp[it(\Gamma_{\mathcal{S}_1\mathcal{E}_k} \pm \Gamma_{\mathcal{S}_2\mathcal{E}_k})]). \tag{7}$$

The calculated decoherence factors in equation (7) will dictate the entanglement behavior of the system. For the case of a homogeneous environment where interaction strengths become $\Gamma_{\mathcal{S}_i\mathcal{E}_k} = \Gamma$, $r_{klmn}(t)$ will take the form of periodic functions [14]

$$\begin{aligned}
r_{\substack{0010\\0111}} &= r^*_{\substack{0100\\1101}} = e^{-it(\omega_{\mathcal{S}_1} \pm \omega_{\mathcal{S}_1\mathcal{S}_2})} \sum_{k=0}^{N} \binom{N}{k} |\alpha|^{2k} |\beta|^{2(N-k)} \exp[i\Gamma(N-2k)t],\\
r_{\substack{0001\\1011}} &= r^*_{\substack{0100\\1101}} = e^{-it(\omega_{\mathcal{S}_2} \pm \omega_{\mathcal{S}_1\mathcal{S}_2})} \sum_{k=0}^{N} \binom{N}{k} |\alpha|^{2k} |\beta|^{2(N-k)} \exp[i\Gamma(N-2k)t],\\
r_{0011} &= r^*_{1100} = e^{-it(\omega_{\mathcal{S}_1} + \omega_{\mathcal{S}_2})} \sum_{k=0}^{N} \binom{N}{k} |\alpha|^{2k} |\beta|^{2(N-k)} \exp[2i\Gamma(N-2k)t],\\
r_{0110} &= r^*_{1001} = e^{-it(\omega_{\mathcal{S}_1} - \omega_{\mathcal{S}_2})}.
\end{aligned} \tag{8}$$

Here, $\omega_{\mathcal{S}_1}$ and $\omega_{\mathcal{S}_2}$ from the system self-Hamiltonians will be canceled out. Thus, the interaction strengths $\omega_{\mathcal{S}_1\mathcal{S}_2}$ and Γ are now the only dominant frequencies for the homogeneous environment. Since both frequencies are $\in \mathbb{Q}*$, this implies that $\omega_{\mathcal{S}_1\mathcal{S}_2}/\Gamma$ is commensurate. Therefore, the periodic behavior of the entangled system is induced by the lower frequency between $\omega_{\mathcal{S}_1\mathcal{S}_2}$ and Γ. Cucchietti et al. [6] suggested that the decoherence in a homogeneous environment will result in a periodic function and approximately Gaussian envelope where the width is inversely proportional to N, as shown in figure 2. Contrary to the results of Sese et al. [8], the recurrence time is independent of the parity of N.

In more realistic scenarios where the environment is usually heterogeneous, the interaction strength $\Gamma_{\mathcal{S}_i\mathcal{E}_k}$ for the kth environment is spontaneously random and hard to control. It was found that an initially sharp concurrence will follow a convoluted entanglement after time t for small N. However, for larger N, similar to the results of [9], sudden death starts to become more noticeable. The entanglement dissipation due to its interaction to heterogeneous environment is similarly observed by [9]. This is due to the fact that we considered a uniformly random $\Gamma_{\mathcal{S}_i\mathcal{E}_k}$ in the range of $[0, 1/2]$, such that the cardinality of $\mathbb{Q}'\dagger$ is much greater than $\mathbb{Q}$. Thus, $\omega_{\mathcal{S}_1\mathcal{S}_2}/\Gamma_{\mathcal{S}_i\mathcal{E}_k}$ is more likely incommensurate. The incommensurability of the frequencies in the sum of periodic functions implies that there will be no common period [15] which will make the concurrence irreversible.

∗Set of rational numbers
†Set of irrational numbers

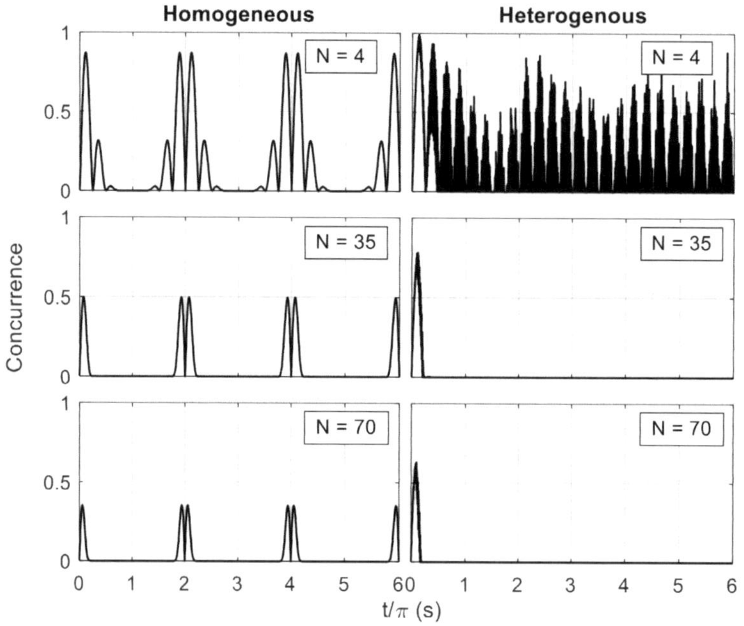

Figure 2: Concurrence for homogeneous (left) and heterogeneous (right) environment. The parameters used are $\omega_{\mathcal{S}_1} = 3$ Hz, $\omega_{\mathcal{S}_2} = 2$ Hz, $\omega_{\mathcal{S}_1\mathcal{S}_2} = 4$ Hz, and $|\alpha|^2 = |\beta|^2 = 1/2$. The environment interaction strength for homogeneous is $\Gamma = 1/2$ Hz. Here, $\Gamma_{\mathcal{S}_i\mathcal{E}_k}$ values for heterogeneous case are taken from uniformly random numbers within $[0, 1/2]$ Hz using Markovian condition.

4. Conclusion

We have shown the entanglement dynamics for initially an separable two-qubit system interacting with an N-qubit environment. For a homogeneous environment, the concurrence of the system behave periodically with its width and periodicity determined from the initial parameters N, $\omega_{\mathcal{S}_1\mathcal{S}_2}$, and $\Gamma_{\mathcal{S}_i\mathcal{E}_k}$. On the other hand, a heterogeneous environment results to an entanglement sudden death, as also predicted by [9, 10]. The entanglement asymptotic decay becomes more pronounced for a larger N qubit environment. This behavior of a bipartite system hints that if we could control the environment, we could influence its entanglement decay rate.

For future work, from a two-qubit system, one could consider a larger N system interacting with an N-qubit environment. One could also utilize chained environments where each qubit environment interacts with its nearest neighbor. It would be interesting to know what would happen if every qubit environment interacted with all the constituent subsystems.

References

[1] E. Schrödinger, Discussion of probability relations between separated systems, *Math. Proc. Cambridge Philos. Soc.* **31**, 555–563 (1935).

[2] R. Horodecki, P. Horodecki, M. Horodecki, and K. Horodecki, Quantum entanglement, *Rev. Mod. Phys.* **81**, 865 (2009).

[3] J. H. Huang and S. Y. Zhu, Sudden death time of two-qubit entanglement in a noisy environment, *Opt. Commun.* **281**, 2156 (2008).

Journal of Physics: Conference Series **2793** (2024) 012015 doi:10.1088/1742-6596/2793/1/012015

[4] W. Gen-Fang, L. Jian-Mei, and W. Lin-Cheng, Spin-star environment assisted entanglement generation in weakly coupled bipartite systems, *Chin. Phys. B* **23**, 050311 (2014).

[5] J. Ying-Hua, L. Yong-Mei, and W. Zi-Sheng, Avoiding the decay of entanglement for coupling two-qubit system interacting with a non-markov environment, *Chin. Phys. B* **20**, 070304 (2011).

[6] F. M. Cucchietti, J. P. Paz, and W. H. Zurek, Decoherence from spin environments, *Phys. Rev. A* **72**, 052113 (2005).

[7] Z. Merali, Quantum computing: The power of discord, *Nature* **474**, 24 (2011).

[8] L. J. F. Sese and E. A. Galapon, Exact recurrence of a qubit immersed in a homogeneous n-qubit environment, in *Proceedings of the Samahang Pisika ng Pilipinas* (2022), vol. 40, SPP-2022-1D-02.

[9] R. Tahira, M. Ikram, T. Azim, and M. S. Zubairy, Entanglement dynamics of a pure bipartite system in dissipative environments, *J PHYS B-AT MOL OPT* **41**, 205501 (2008).

[10] E. G. El-Hadidy, A.-H. Abdel-Aty, and A. Farouk, New properties of a pure bipartite system in presence of dissipative environments, *Alex. Eng. J.* **59**, 1215 (2020).

[11] M. Flores and E. Galapon, Two qubit entanglement preservation through the addition of qubits, *Ann. Phys.* **354**, 21 (2015).

[12] Z. Gedik, Spin bath decoherence of quantum entanglement, *Solid State Commun.* **138**, 82 (2006).

[13] W. K. Wootters, Entanglement of formation of an arbitrary state of two qubits, *Phys. Rev. Lett.* **80**, 2245 (1998).

[14] H. Bohr, *Almost Periodic Functions* (Chelsea Publishing Company, New York, United States of America, 1947).

[15] C. Cordunneanu, *Almost periodic functions* (American Mathematical Society, Providence, RI, 1989).

A survey on topological solitons in planar nonlinear Dirac models

R Dikole

Optronic Sensor Systems, Defence and Security, Council for Scientific and Industrial Research, South Africa

Email address: rdikole@csir.co.za

Abstract. In this text, we review tardyonic and tachyonic planar Dirac models with several cubic nonlinearities such as the Kerr, Soler and massive Thirring nonlinearities. These models have relevance in a newly discovered class of solids named topological insulators and have been shown to exhibit topological properties such as the Berry phase and Chern number. Moreover, nonlinear Dirac models have vortex solutions and edge state solitons. As understood in quantum mechanics that a Hamiltonian must yield real eigenvalues or dispersion relation, we argue that Dirac equations in the tachyonic case are valid when studied in non-vanishing backgrounds because this gets rid of the complexity in the dispersion relation.

1. Introduction

The Dirac equation bears relevance today due to its connection with graphene [1] and a newly discovered class of solids known as topological insulators. These are materials which exhibit special scattering resisting states [2] on the surfaces called topological edge states [2, 3]. Topological edge states are realisable in condensed matter electronic systems, and photonic crystals in optics [4, 5], they have also been predicted in a large number of photonic systems including arrays of coupled optical resonators, helical waveguide arrays and gyromagnetic photonic crystals [2]. Topological edge states bear the name "topology" because they possess topological invariants, these are quantities that remain unaffected by continuous deformation of an object. Examples of topological invariants in lattice models include the Berry phase and Chern number.

Topological invariants can be used to describe band structures in certain periodic crystalline materials [2]. The simplest model that exhibits topological properties is the Su-Schrieffer-Heeger model which describes a one-dimensional dimer chain with alternating strong and weak nearest-neighbour couplings $\kappa_{1,2}$. The model was introduced by Su, Schrieffer and Heeger to study soliton formation in polyacetylene [6]. The Su-Schrieffer-Heeger Hamiltonian in Fourier space is a single-particle Bloch Hamiltonian [7] of the form

$$H_{SSH} = -(\kappa_1 + \kappa_2 cos(kd))\sigma_1 - \kappa_2 sin(kd)\sigma_2, \tag{1}$$

where $\sigma_{1,2}$ are the Pauli spin matrices and d is the lattice spacing. This Hamiltonian (1) reduces to the one-dimensional Dirac Hamiltonian in the vicinity of the Brillouin zone edge $kd = \pi$ [7]. By Brillouin zone, it is meant the primitive cell in reciprocal space i.e. the Fourier transform of real space. The two energy bands $\mathcal{E}_{\pm}$ are separated by a gap when $\kappa_1 \neq \kappa_2$ with a band closure occurring at $\kappa_1 = \kappa_2$ [2, 8].

Edges states of the Su-Schrieffer-Heeger model occur in the middle of the bandgap when the coupling is weak, that is $\kappa_2 < \kappa_1$. These edge states have a localisation in one of the sublattices and decay

Journal of Physics: Conference Series **2793** (2024) 012016 doi:10.1088/1742-6596/2793/1/012016

exponentially far away from the edge of the lattice. The edge states are topologically protected in the sense that their frequency is fixed to zero and they cannot be destroyed by chiral symmetry-preserving perturbations. Edge states also emerge in two-dimensional lattices such as hexagonal lattices, which are two-dimensional generalisations of the Su-Schrieffer-Heeger lattices. These include photonic graphene, staggered graphene and the Haldane model. Under consideration of nonlinear effects like the Kerr nonlinearity, for example, these planar models can exhibit topological solitons and vortices, with photonic graphene being the classic example.

The aim of this paper is to provide an overview of two-dimensional nonlinear continuous wave equations in the and their associated solutions, in particular, vortex solutions. This paper provides the reader with a summary of such two-dimensional models exhibiting vortex solution and other topological phenomena such as edge state solitons. We consider the works of leading experts such as Haldane, Smirnova, Barashenkov and Comech *et al.*

This paper is organised as follows: section 2 presents the method for finding vortex solutions, section 3 presents another nonlinear Dirac-type equation with cubic nonlinearity, namely, the tachyonic Soler model. This is a Soler model whose linear part is such that the spinor components satisfy the imaginary mass Klein-Gordon equation. The standard Soler model is presented in section 3.2. The nonlinear Dirac equation in photonic graphene is presented in section 4. Finally, section 4 provides the discussion and section 5 concludes this paper.

2. Method

The method employed in the search for vortex solutions in the sections that follow involves the assumption of a vortex ansatz $\psi_1 = f_n(r)e^{in\theta}, \psi_2 = g_{n-1}(r)e^{i(n-1)\theta}$, with ψ_1 and ψ_2 being the components of a $(2+1)$ spinor and $f_n(r)$ and $g_{n-1}(r)$ are functions of the radial coordinate, while the complex exponential accounts for the radial symmetry of the vortex. This solution form reduces the nonlinear Dirac equation to systems of coupled nonlinear ordinary differential equations which can be solved via a numerical routine.

3. The tachyonic Soler model

The Soler model is a nonlinear Dirac-type equation with a scalar self-interaction potential of the form $\left(\overline{\psi}\psi\right)^2$. In $(2+1)$dimensions Soler nonlinearity coincides with the Thirring nonlinearity as shown in Ref. [9] because of the equality of the γ^5 matrix with the 2×2 identity matrix. Note that the imaginary mass Dirac equation (tachyonic) can be found either by mapping $m \mapsto im$ or multiplication of the mass term with γ^5 in the Lagrangian. Because of this equality, the Soler model presented here is deemed to be non-Lagrangian. Therefore, the $(2+1)$ dimensional tachyonic Soler model takes the form

$$\begin{aligned} u_t + (\partial_x + i\partial_y)v - u(|u|^2 - |v|^2) = 0, \\ v_t + (\partial_x - i\partial_y)u - v(|u|^2 - |v|^2) = 0, \end{aligned} \tag{2}$$

The Dirac γ-matrices are chosen as $\gamma^0 = \sigma^3, \gamma^1 = i\sigma^2$ and $\gamma^2 = -i\sigma^1$. This model is analogous to a generalisation of the spinor extension of the complex sine-Gordon model. To put this into perspective, consider a slight modification of (2):

$$\begin{aligned} u_t + (\partial_x + i\partial_y)v - mu - u(A_{11}|u|^2 - A_{12}|v|^2) = 0, \\ v_t + (\partial_x - i\partial_y)u + mv - v(A_{21}|u|^2 - A_{22}|v|^2) = 0, \end{aligned} \tag{3}$$

where $A_{ij}$are real constants, $i,j = \{1,2\}$ and m is the mass. The $(2+1)$ dimensional extension of the Euclidean massive Thirring model presented in Ref. [10] can be obtained by letting $m = 1$, $A_{11} = A_{22} = 0$ and $A_{12} = A_{21} = 1$, which of course admits vortex solutions. The model in question (2) is obtained by letting $m = 0$ and $A_{ij} = 1$, with $i,j = \{1,2\}$. The choice $m = 0$ is to avoid the complex roots in the constant background dispersion relation. It remains to be found whether this version of the Soler model (2) admits stationary vortex solutions. Since this model is tachyonic this means that its zero-background dispersion relation has complex roots and thus cannot have any physical interpretation. To revert this to physics, as in the case of the complex sine-Gordon model, the dispersion relation is best considered in a nonvanishing background. As in the case of the complex sine-Gordon, the system in question (2) admits a constant solution of the form $u = 1$ and $v = e^{i\alpha}$. Expanding (2) around this solution yields

$$\dot{\delta u} + (\partial_x + i\partial_y)\delta v + (\delta\overline{u} + \delta u - (e^{i\alpha}\delta\overline{v} + e^{-i\alpha}\delta v)) = 0,$$
$$\dot{\delta v} + (\partial_x - i\partial_y)\delta u - e^{i\alpha}(\delta\overline{u} + \delta u - (e^{i\alpha}\delta\overline{v} + e^{-i\alpha}\delta v)) = 0. \quad (4)$$

Letting $\delta u = a + ib$, $\delta v = p + iq$ and splitting the resulting equations into the real and imaginary parts and considering waves of the form $(a, b, p, q)^T = (A, B, P, Q)^T e^{i(\omega t - \boldsymbol{k}\cdot\boldsymbol{x})}$ results in a polynomial of the form:

$$(\omega^2 - |\boldsymbol{k}|^2)^2 = 0, \quad (5)$$

this quantity is always positive, implying that its roots are always real. On the contrary, the choice $m = 0$, $A_{11} = A_{12} = 1$,and $A_{21} = A_{22} = -1$ yields another Soler-type equation whose non-vanishing background dispersion relation takes the form:

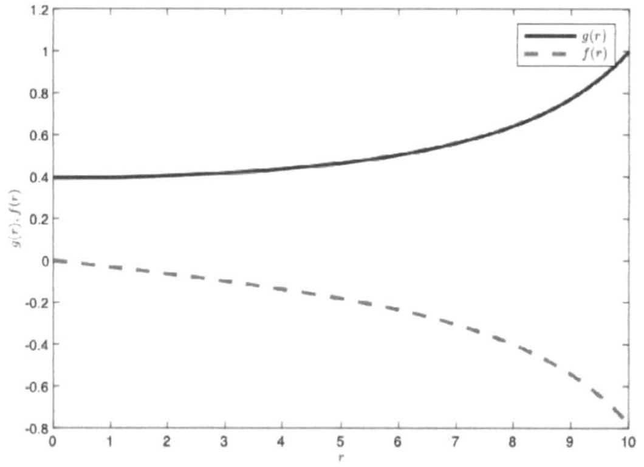

Figure 1. The numerical solution of single vortex ordinary differential equation (10) obtained via a nonlinear shooting method with boundary conditions $f(r) \to 0$ as $r \to 0$ and $f \to 1$ as $r \to \infty$. The solution is computed on the interval $r \in [0,10]$.

$$\omega^2 + 4i\omega - |\boldsymbol{k}|^2 + 4i(k_1 sin\alpha - k_1 cos\alpha) = 0. \quad (6)$$

This equation of course has complex valued roots, which implies that this particular choice of A_{ij} does not produce a system with a physical interpretation.

3.1. Stationary vortices in the tachyonic Soler model

To search for stationary vortex solutions, we set the time derivatives to zero and write (2) in polar coordinates:

$$e^{-i\theta}(\partial_r - ir^{-1}\partial_\theta)v - u(|u|^2 - |v|^2) = 0$$
$$e^{i\theta}(\partial_r + ir^{-1}\partial_\theta)u - v(|u|^2 - |v|^2) = 0. \quad (7)$$

We are looking for solutions of the form

$$u = f_n(r)e^{in\theta}, v = g_{n-1}(r)e^{i(n-1)\theta}, \quad (8)$$

where the f_n and g_{n-1} are real valued functions of the radial coordinate r and n is the vorticity. Substituting solution (8) into (7) yields

$$g'_{n-1} - (n-1)r^{-1}g_{n-1} - f_n(f_n^2 - g_{n-1}^2) = 0,$$
$$f'_n + nr^{-1}f_n - g_{n-1}(f_n^2 - g_{n-1}^2) = 0, \quad (9)$$

where the prime denotes the first derivative. The single vortex equations are obtained by setting the vorticity to unity ($n = 1$) in the above system, that is

$$g' - f(f^2 - g^2) = 0,$$
$$g' + r^{-1}f - g(f^2 - g^2) = 0. \quad (10)$$

The numerical solution to the single vortex differential equations (10) can be sought by a nonlinear shooting method for two-point boundary value problems where a nonlinear boundary value problem is converted into a series of initial value problems. This method is fully described in the eleventh chapter of Ref. [10]. The boundary conditions are $f(r) \to 0$ as $r \to 0$ and $f \to 1$ as $r \to \infty$ or $g'(r) \to 0$ as $r \to 0$ and $g'(r) \to 0$ as $r \to \infty$. Fig. 1 shows the output obtained from the nonlinear shooting method, with both f and g growing for large values of r and thus suggesting that this version of the Soler model (2) does not admit vortex solutions.

3.2. Stationary vortices in the tardyonic Soler model

The standard Soler model, that is, the Soler model associated with the tardyonic Dirac equation has been reported to admit stationary vortex solutions by Cuevas-Maraver *et al* [8]. The authors of Ref. [8] present a nonlinear Dirac equation with a generalised Soler-type nonlinearity with a Lagrangian density of the form:

$$\mathcal{L} = \overline{\psi}(i\gamma^\mu \partial_\mu - m)\psi + 0.5\left(\overline{\psi}\psi\right)^{k+1}(k+1)^{-1}, \mu = 0,1,2,\ k \in \mathbb{N}. \tag{11}$$

The choice $\gamma^0 = \sigma^3$, $\gamma^1 = i\sigma^2$ and $\gamma^2 = -i\sigma^1$, with $k = 1$ and $m = 1$ yields a massive $(2 + 1)$dimensional nonlinear Dirac equation of the form

$$\begin{aligned} &i\partial_t u + e^{-i\theta}(i\partial_r + r^{-1}\partial_\theta)v - u + (|u|^2 - |v|^2)u = 0, \\ &i\partial_t v + e^{i\theta}(i\partial_r - r^{-1}\partial_\theta)u + v - (|u|^2 - |v|^2)v = 0, \end{aligned} \tag{12}$$

written in polar coordinates. Vortex solutions are obtained by substituting a solution ansatz $\psi(t,x) = e^{-i\omega t}\phi(r)$, where $\phi(r) = \left(g(r)e^{in\theta}, if(r)e^{i(n+1)\theta}\right)^T f, g \in \mathbb{R}$, so that one obtains a system of nonlinear vortex ordinary differential equations

$$\begin{aligned} &\omega g - \left(\partial_r + (n+1)r^{-1}\right)f - g + \left(g^2 - f^2\right)g = 0, \\ &\omega f + (\partial_r - nr^{-1})g + f - (g^2 - f^2)f = 0. \end{aligned} \tag{13}$$

Equation (13) is a nonlinear differential equation whose solutions are solitary wave structures, that is localised waves whose localisation shifts towards the origin for large values of the frequency ω. Fig 2 depicts this phenomenon in the $n = 1$ case— the spinors begin with localisation at $r = 3$ for $\omega = 0.2$ and move closer and closer to the origin as ω grows closer to one. These images agree with those of Cuevas-Maraver *et al* [8]— these solitary waveforms were also obtained in the context of photonic graphene with Kerr-like nonlinearities in Ref. [12]. The authors of Ref. [9] report that solitary waves are spectrally when the angular frequency drops below a dimension-dependent critical value $\omega_c \approx 0.936$ and that solutions of higher vorticity are unstable and split into lower vortex charge configurations to preserve the overall vorticity.

4. Nonlinear Dirac equation in photonic graphene

The tight binding approximation of the photonic graphene Hamiltonian, that is, an array of laser beams organised in a hexagonal lattice, near the Dirac points corresponds to the massless Dirac equation. Thus, in optics, the nonlinear extension of the Dirac-like Hamiltonian involves adding a Kerr nonlinearity, which is cubic-type nonlinearity where the spinor components are decoupled with the nonlinearity. Due to this decoupling of spinor components with the nonlinearity, the $(2 + 1)$dimensional nonlinear Dirac equation with Kerr nonlinear is not Lorentz invariant as already shown by [12]. This model is presented in Refs. [7, 12, 13] as:

$$\begin{aligned} &i\partial_t\psi_1 = -i\partial_x\psi_2 - \partial_y\psi_2 + (M - g|\psi_1|^2)\psi_1, \\ &i\partial_t\psi_2 = -i\partial_x\psi_1 + \partial_y\psi_1 - (M + g|\psi_2|^2)\psi_2. \end{aligned} \tag{14}$$

4.1. Topological edge states

This model is known to have topological edge states since it is the $(2 + 1)$ dimensional extension of the $(1 + 1)$ dimensional nonlinear lattice model, the nonlinear Su-Schrieffer-Heeger-Model. The edge states are obtained by restricting oneself to a single axis and thus reducing the problem to a $(1 + 1)$ dimensional problem. A one-dimensional reduction of system (14) is performed by letting $\psi_1(x,y,t) = b(y,t)e^{ikx}$ and $\psi_2(x,y,t) = a(y,t)e^{ikx}$:

$$i\partial_t a = \partial_y b + kb - (M + g|a|^2)a,$$

$$i\partial_t b = -\partial_y a + ka + (M - g|b|^2)b. \tag{15}$$

The system (15) is a $(1 + 1)$ dimensional nonlinear Dirac equation is written in a non-standard form. To write (15) in the standard Dirac gamma matrix representation (spinor) we make the following change of variables

$$u = (b + ia)/2\delta, v = (b - ia)/(2\gamma), \tag{16}$$

and

$$a = i(\gamma v - \delta u), b = \delta u + \gamma v. \tag{17}$$

Using (16)— (17) we can turn (17) into spinor equations as

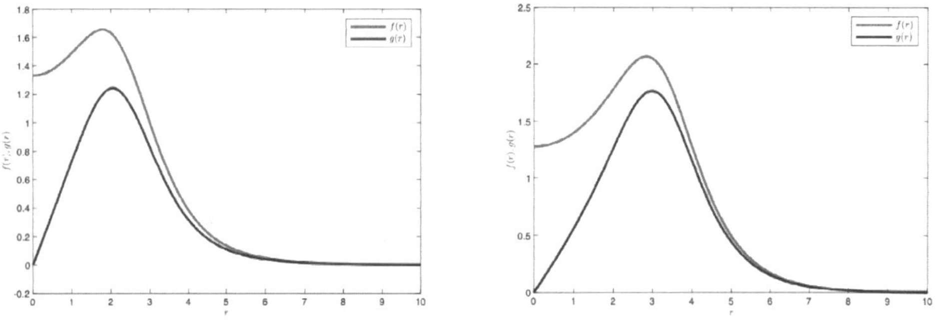

(a) (b)

Figure 2. The numerical solution of the vortex differential equation (13) via the shooting method. The solution is computed for several values of the frequency omega, namely, (a) $\omega = 0.2$, (b) $\omega = 0.3$ The solution is computed with vorticity $n = 1$.

$$\begin{aligned} i(\partial_t u - \partial_y u) + (M + ik)v + \mathcal{F}(u, v)v + \mathcal{G}(u, v)u = 0, \\ i(\partial_t v + \partial_y v) + (M - ik)v + \mathcal{F}(u, v)u + \mathcal{G}(u, v)v = 0, \end{aligned} \tag{18}$$

where $\delta = -1$ and $\gamma = 1$ and

$$\mathcal{F}(u, v) = g(uv^* + vu^*), \mathcal{G}(u, v) = g(|u|^2 + |v|^2). \tag{19}$$

Equation (18) is a $(1 + 1)$ dimensional nonlinear Dirac equation written using the standard choice of γ-matrices. The newly transformed nonlinear Dirac equation is not Lorentz invariant because of the term $\mathcal{G}(u, v)$which does not transform like a spinor. The solution to (10) has been obtained by [7] by search for solutions proportional to $e^{-i\omega t}$ and applying the Lorentz transformation $\tau = \gamma(t - vy)$, $\xi = \gamma(y - vt)$ with $\gamma = \frac{1}{\sqrt{1-v^2}}$. The solution takes the form

$$(u, v)^T = \sqrt{2\rho_s(\xi)}e^{-i\omega t}(cos\alpha_s(\xi), -sin\alpha_s(\xi))^T \tag{20}$$

where ρ_s is the intensity and α_s is the spin angle. The exact expressions of ρ_s and $\alpha_s$are presented in Ref. [7].

4.2.Stationary vortices in photonic graphene

To find the vortex solution of (14) it is useful to cast it into its polar coordinate form as follows:

$$\begin{aligned} i\partial_t\psi_1 = -e^{-i\theta}(i\partial_r + r^{-1}\partial_\theta)\psi_2 + (M - |\psi_1|^2)\psi_1, \\ i\partial_t\psi_2 = -e^{i\theta}(i\partial_r - r^{-1}\partial_\theta)\psi_1 - (M + |\psi_2|^2)\psi_2. \end{aligned} \tag{21}$$

The vortex ansatz in the case of the nonlinear Dirac equation with Kerr nonlinearity is takes the form:

$$(\psi_1, \psi_2)^T = \left(fe^{in\theta}, ige^{i(n+1)\theta}\right)^T e^{-i\omega t}, \tag{22}$$

where $n = 0, \pm 1, \pm 2, \ldots$ is the azimuthal number or simply vorticity. Solutions of the form (22) can be interpreted as radially symmetric two-dimensional bulk solitons. Substituting this ansatz into (21) yields a system of spinor ordinary differential equations

$$\begin{aligned} g' + (n + 1)r^{-1}g + (M - \omega - f^2)f = 0, \\ f' - nr^{-1}f + (M + \omega + g^2)g = 0. \end{aligned} \tag{23}$$

The numerical solution for this system and its stability analysis has been presented in Ref. [13]. These are shown in Fig. 3 where the vorticity is fixed to $n = 1$ and the numerical solution is computed for

Journal of Physics: Conference Series **2793** (2024) 012016 doi:10.1088/1742-6596/2793/1/012016

several values of ω, namely $\omega = 0.3$, $\omega = 0.5$, with $M = 1$ and boundary conditions on $g(r)$ are $g(0) = 0$ and $g(\infty) = 0$. On the other hand, the boundary conditions on $f(r)$ are such that $f(\infty) = 0$, with the best possible starting value of $f(0)$ being computed with Newton's method or *MATLAB*'s root finder `fsolve()`. These solutions are stable for values of ω above the critical angular frequency $\omega_c > 0.338$ [13]. The effect of ω on the radial solution is shown in Fig. 3 where the amplitude of the localised spinors decreases with an increase in omega, with the width of the localised solutions widening. The radial solution starts with a localisation near the origin when $\omega = 0.3$, the localisation then moves further away from the origin with a rise in ω.

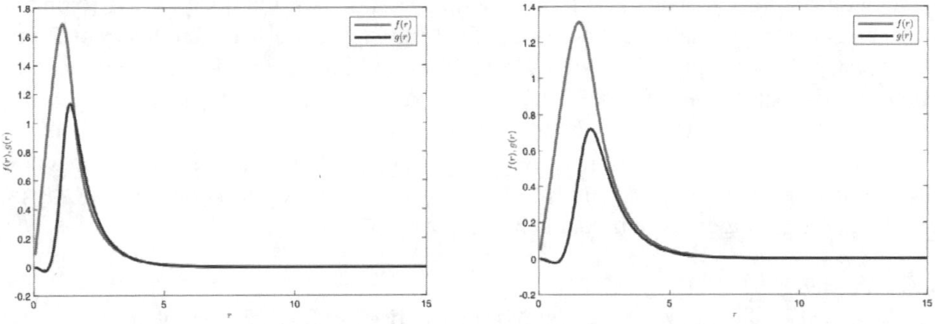

(a) (b)

Figure 3. The numerical solution of the nonlinear vortex differential equation (21) via the shooting method. The solution is computed for several values of the frequencyω, namely, (a) $\omega = 0.3$, (b) $\omega = 0.5$. The solution is computed with vorticity $n = 1$, and $M = 1$.

5. Discussion

We have looked at $(2+1)$dimensional nonlinear Dirac equations with cubic nonlinearities; these include a generalisation of the Euclidean massive Thirring model presented in Ref. [9] and it was identified as the tachyonic Soler model. The authors of Ref. [9] have shown that the Euclidean massive Thirrring model exhibits topological solitons. It can be argued that although the model in Ref. [9] possesses a complex-valued dispersion relation in vanishing backgrounds thus making it nonphysical, the physics of the model can be recovered by studying it in nonvanishing backgrounds, i.e. the constant solution $u = 1$, $v = e^{i\alpha}$.

Similarly, the complexity of the dispersion relation in Eq. (3) can be rid by linearising Eq. (3) about the solution $u = 1, v = e^{i\alpha}$. This model, however, was found to not have any vortex formations in the case of vorticity $n = 1$. The model was also found to be non-Lagrangian due to the coincidence of the γ^5 matrix with the 2×2 identity matrix. The γ^5 matrix bears relevance because the imaginary mass Dirac equation can be found either by mapping $m \mapsto im$ or multiplication of the mass term with γ^5 in the Lagrangian. We also looked at the standard Soler model, where the word "standard" refers to the fact that this is a nonlinear Dirac equation whose spinor components satisfy the Klein-Gordon equation with real mass i.e. a Lorentz factor which is always real, where v is a velocity not to be confused with the spinor component.

The standard Soler model was shown by [8] to have vortex solutions - the single vortex solution is shown in Fig. 2, where the effects of the angular frequency ω were shown to shift the solitary wave localisation as ω increases. These solitary waves were reported to be spectrally when the angular frequency drops below a dimension-dependent critical value $\omega_c \approx 0.936$ by the authors of Ref. [13]. The aforementioned authors also report that solutions of higher vorticity are unstable and split into lower vortex charge configurations to preserve the overall vorticity.

Finally, we looked at the nonlinear Dirac equation with Kerr nonlinearity. This model is a description of the propagation of light in lasers arranged in a hexagonal lattice. The symmetry properties of Eq. (14)

were considered by Haddad *et al* [12] and it was found the model is not Lorentz invariant. The solitary wave solutions were also presented in Ref. [13] with their stability existing only for $\omega_c > 0.338$.

6. Conclusion

This paper studied planar nonlinear Dirac equations which have been foundational in the discovery of Dirac materials i.e. a new class of solids called topological insulators and the solutions they possess called topologically protected states and topological edge states. In the context of photonic systems, topological phases can be modelled by nonlinear tight-binding models or nonlinear continuous wave equations. The has been a number of theoretical predictions of nonlinear phenomena in photonic, these include solitons, edge states and bulk solitons. In this paper we reviewed planar topological solitons called vortices in nonlinear Dirac with several nonlinearity i.e. Kerr, Soler and Thirring-type nonlinearities, these were radially symmetric waves that approach zero at the origin and a constant as the radial coordinate goes to infinity. Vortex solutions were found in all models except the tachyonic Soler model, thus suggesting that the model is missing a term.

References

[1] Castro Neto, A. H., Guinea, F., Peres, N. M. R., Novoselov, K. S. & Geim, A. K. The electronic properties of graphene. *Rev. Mod. Phys.* **81**, 109–162 (2009)

[2] Smirnova, D., Leykam, D., Chong, Y. & Kivshar, Y. Nonlinear topological photonics. *Applied Physics Reviews* **7**, 021306 (2020).

[3] Hasan, M. Z. & Kane, C. L. Colloquium: Topological insulators. *Rev. Mod. Phys.* **82**, 3045–3067 (2010).

[4] Ozawa, T. *et al.* Topological photonics. *Rev. Mod. Phys.* **91**, 015006 (2019).

[5] Xie, B.-Y. *et al.* Photonics meets topology. *Opt. Express* **26**, 24531–24550 (2018).

[6] Su, W. P., Schrieffer, J. R. & Heeger, A. J.Solitons in polyacetylene. *Phys. Rev. Lett.* **4**, 1698–1701 (1979).

[7] Smirnova, D. A., Smirnov, L. A., Leykam, D. & Kivshar, Y. S. Topological edge states and gap solitons in the nonlinear dirac model. *Laser & Photonics Reviews* **13,** 1900223 (2019)

[8] Cuevas–Maraver, J., Kevrekidis, P. G., Saxena, A., Comech, A. & Lan, R. Stability of solitary waves and vortices in a 2d nonlinear dirac model. *Physical Review Letters* **116** (2016).

[9] Barashenkov, I. V. & Getmanov, B. S. Multisoliton solutions in the scheme for unified description of integrable relativistic massive fields. non-degenerate $sl(2, \mathbb{C})$ case. *Communications in Mathematical Physics* **112**, 423–446 (1987).

[10] Rosen, D., Burden, R. L., Faires, J. D. & Reynolds, A. C. Numerical analysis. *The American mathematical monthly* **87**, 231– (1980).

[11] Poddubny, A. N. & Smirnova, D. A. Ring Dirac solitons in nonlinear topological systems. *Phys. Rev.* A **98**, 013827 (2018).

[12] Haddad, L. & Carr, L. The nonlinear dirac equation in Bose–Einstein condensates: Foundation and symmetries. *Physica D: Nonlinear Phenomena* **238**, 1413–1421 (2009).

[13] Cuevas-Maraver, J., Kevrekidis, P. G., Aceves, A. B. & Saxena, A. Solitary waves in a two-dimensional nonlinear Dirac equation: from discrete to continuum. *Journal of Physics A: Mathematical and Theoretical* **50**, 495207 (2017).

Journal of Physics: Conference Series **2793** (2024) 012017 doi:10.1088/1742-6596/2793/1/012017

Predicting the minimum energy pathway of 1H to 1T phase transition of select 2D transition metal dichalcogenides via density functional theory and machine learning approach

KY M Diego, D B Putungan and A B Santos-Putungan*

Institute of Mathematical Sciences and Physics, University of the Philippines Los Baños, Philippines

* Corresponding author's email: absantos1@up.edu.ph

Abstract. This study predicts the minimum energy pathways (MEPs) and transition barriers of two-dimensional transition metal dichalcogenides (TMDs) during their 1H to 1T structural phase transitions. The investigation utilizes density functional theory (DFT) calculations and machine learning algorithms to predict the MEPs and transition barriers. Six TMDs, namely $NbSe_2$, ScS_2, $ScSe_2$, $TiTe_2$, VS_2, and VSe_2, are selected for analysis. The DFT calculations provide reference values for comparison with the machine learning predictions. The transition barriers obtained through DFT calculations range from 0.13 eV to 0.72 eV. The machine learning models, including Linear Regression, Random Forest Regression, K-Nearest Neighbor Regression, and Gaussian Process Regression, demonstrate varying degrees of accuracy in predicting the transition barriers. Gaussian Process Regression achieved the highest accuracy among all ML models, with the absolute mean error spanning all systems being 0.0093. It has been shown that the features that were identified in the study are sufficient for the prediction of the MEPs and transition barriers of TMDs using machine learning. Overall, this study contributes to the understanding of TMD phase transitions and provides insights into the integration of machine learning in materials science research.

1. Introduction

Two-dimensional transition metal dichalcogenides (TMDS) have received substantial interest in recent times due to their distinctive and versatile properties, including the ability to exhibit semiconducting to superconducting behavior [1-2], tuneable direct bandgaps [3], and a broad range of thermal conductivity [4]. These are also known to have excellent near-infrared absorbance [5], and sensitivity to thickness [6-7], Moreover, many properties of TMD can be tuned and improved using phase transition [8]. The unique combination of electronic, optical, and mechanical characteristics makes TMDS an attractive option for various technological applications, such as photovoltaics, electronic devices, and catalysts. TMD monolayers largely occur in either 1H or 1T structural phases [9]. These arrangements influence the compounds' electronic properties, leading to tunable bandgaps, high carrier mobility, and perfect electron-hole symmetry. The investigation and analysis of the MEPs and transition barriers of 2D TMDs during their 1H to 1T structural phase transitions hold significant implications in materials science.

To better understand these structures and their applications, researchers have employed various computational methods, such as the density functional theory (DFT). However, DFT computations are computationally expensive and may become time-consuming and costly for large-scale studies. The

Journal of Physics: Conference Series **2793** (2024) 012017 doi:10.1088/1742-6596/2793/1/012017

advancement of technology has led to the integration of machine learning techniques into the study of materials.

Machine learning has proven to be an indispensable tool in material science due to its ability to identify trends and make predictions based on an existing dataset [10]. This potentially reduces the computational cost and time required to study materials, making it a more efficient alternative to traditional density functional theory (DFT) computations. This study aims to utilize the machine learning approach to predict the minimum energy pathway of TMDs from the 1H to 1T structural phase. DFT computations are used to form a training dataset for the machine learning model, which is subsequently used to predict the energies of each TMD structure using its parameters. The prediction accuracy of the model is then tested, exhibiting the effectiveness of the machine learning approach in materials science.

2. Methodology

All density functional theory-based computations were executed in Quantum Espresso [11]. A cut-off energy of 80 Ry was used in plane wave expansion, and a $4 \times 4 \times 1$ Monkhorst-Pack k-point mesh was used in Brillouin zone sampling. Projector-augmented wave (PAW) pseudopotentials were used in the calculations and the exchange-correlation energy was approximated through the vdW-DF functional using the revised Perdew-Burke-Ernzerhof (rev-PBE) functionals. The $(1 \times 1 \times 1)$ unit cells of the 1H and 1T structural phases of six transition metal dichalcogenides were prepared for the DFT computations. The analysed TMDs include $NbSe_2$, ScS_2, $ScSe_2$, $TiTe_2$, VS_2, and VSe_2. A 15-Å vacuum layer was added in the *z*-direction of each structure to prevent interfacial interactions from repeat cells. The nudged elastic band (NEB) method calculations were used to produce fifty images for each TMD, which mapped out their theoretical minimum energy pathways. The transition barriers of each DFT-generated TMD MEP were taken from the activation energy spanning from the local minimum at around the 1H phase to where the transition state occurs in the MEP. Numerically, this would be calculated as

$$\Delta G = G_{TS} - G_R$$

where G_{TS} is the total energy of the transition state and G_R is the total energy of the reactant.

The database consists of 300 data points, which were split into training and testing databases in a ratio of 80% - 20%, with the testing database consisting of the transition barrier for each structure along with nine other surrounding points in the potential energy surface. The Scikit- learn package was imported from Python for scaling of data and the creation and optimization of machine learning models for the TMD database. Supervised learning was implemented for this study. Various models were fitted, including Linear Regression (LR), Random Forest Regression (RFR), Gaussian Process Regression (GPR), and K-Nearest Neighbor Regression (KNN). The hyperparameters for each model were chosen based on their resulting accuracy after testing. Among the machine learning models used, only Linear Regression is parametric. Random Forest Regression, K-Nearest Neighbor Regression, and Gaussian Process Regression are non-parametric, as the model structure is determined solely by the given data.

3. Results and discussion

3.1 DFT-calculated minimum energy pathways of TMDs

Figure 1a depicts the nearly symmetrical minimum energy pathway of $NbSe_2$ from the 1H to 1T structural phase transition. The transition phase of the structure is observed to have occurred at -15483.65 Ry, indicating the energy level at which the transition occurs. The calculated transition barrier for the structure is 0.72 eV, the highest among all selected TMDs. This implies that a relatively higher amount of energy is required for a phase transition to occur. It is also observed that 1H-$NbSe_2$ is more stable than 1T-$NbSe_2$ by 0.09 eV.

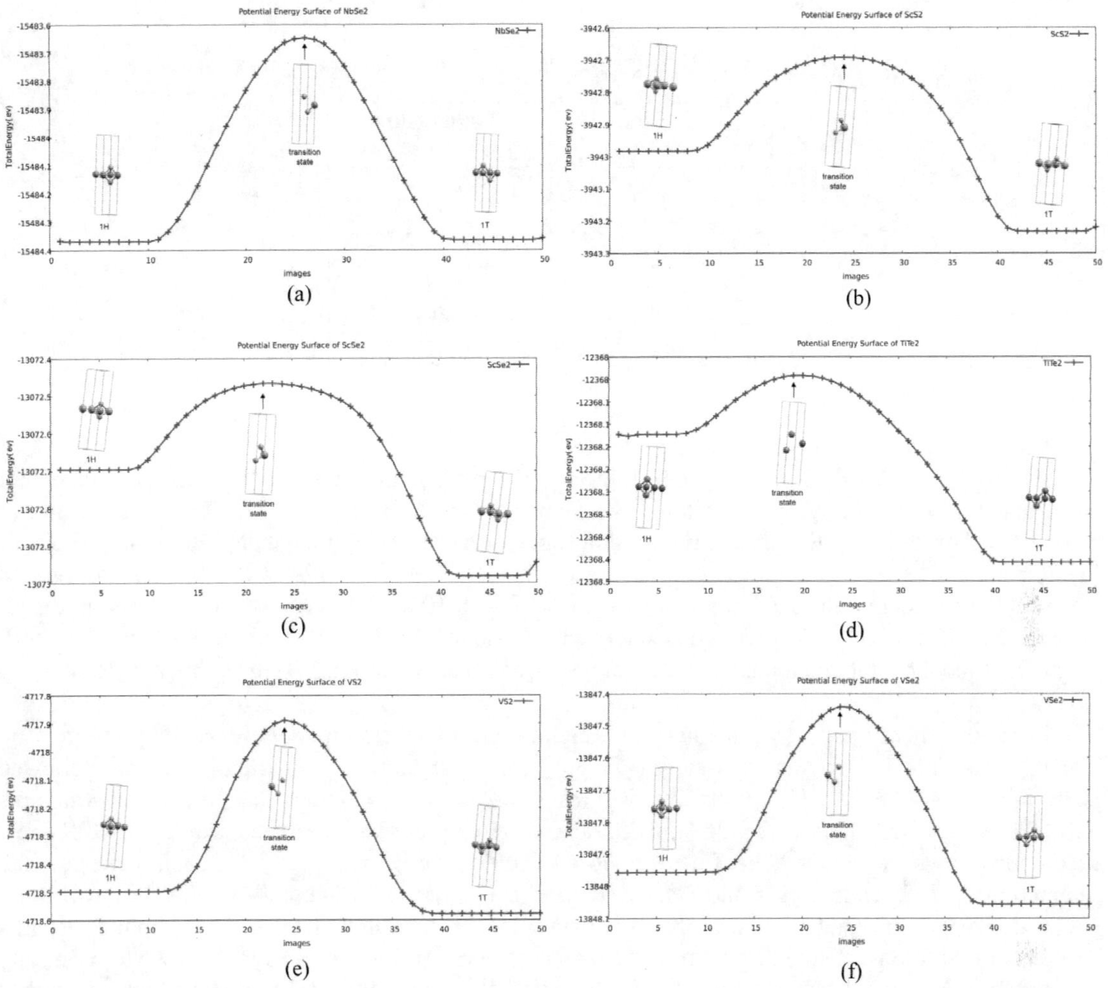

Figure 1. Potential Energy Surfaces (PES) from 1H to 1T structural phase of (a)$NbSe_2$, (b) ScS_2, (c) $ScSe_2$, (d) $TiTe_2$, (e) VS_2, and (f) VSe_2.

The minimum energy pathways of ScS_2 and $ScSe_2$ are illustrated in Figures 1b and 1c, respectively. Both structures exhibit similar behaviors, with the 1T phases considerably more stable than their 1H counterparts. For ScS_2, a transition barrier of 0.29 eV was obtained, while $ScSe_2$ exhibited a slightly lower transition barrier of 0.23 eV.

Figure 1d exhibits the minimum energy pathway of $TiTe_2$ from the 1H to 1T structural phase transition. The total energy of 1T-$TiTe_2$ is lower than that of 1H-TiTe2, indicating a more stable structure in the 1T phase. Remarkably, $TiTe_2$ exhibits the lowest calculated transition barrier among the six studied 2D TMDs, with a value of 0.13 eV. This suggests that the phase transition from 1H to 1T in $TiTe_2$ requires relatively less energy compared to the other TMDs. The minimum energy pathways of VS_2 and VSe_2 are depicted in Figures 1e and 1f, respectively. Similar to ScS_2, $ScSe_2$, and $TiTe_2$, both VS_2 and VSe_2 exhibit more stable 1T structures compared to their 1H structures. For VS_2, the calculated transition barrier is 0.61 eV. This transition barrier suggests that a significant amount of energy is needed to achieve the transformation. Similarly, VSe_2 demonstrates a transition barrier of 0.52 eV, implying a considerable energy cost for the phase transition. Results show the different transition barrier values obtained for each 2D TMD. Further analysis of these values would assist in their precise control and tunability, proving beneficial in the fields of catalysis and energy storage among other applications.

Table 1. Transition Barriers of TMDs via DFT-Based Calculations.

TMDs	**Transition Barrier, ΔG (eV)**
$NbSe_2$	0.718
SCS_2	0.291
$ScSe_2$	0.229
$TiTe_2$	0.132
VS_2	0.608
VSe_2	0.517

3.2 Predicted minimum energy pathways

The average mean errors of the machine learning algorithm predictions were calculated to evaluate their performance. Linear Regression exhibited the highest absolute mean error (AME), with a value of 1684.1028. This indicates that Linear Regression is not an accurate fit for the data, likely due to the non-linear behavior of the minimum energy pathway curves. Following this is the decision tree-based Random Forest Regression (RFR) model, with an AME of 0.1072. A lower AME is observed in the K-Nearest Neighbor (KNN) Regression Model, with a value of 0.0707. The Gaussian Process (GP) Regression model exhibited the highest accuracy among all tested models, with an AME of 0.0093.

The minimum energy pathway plots generated by the GP Regression model are shown to be in good general agreement with the density functional theory-based calculations. In particular, the plots showed near alignments with the DFT results in structures $ScSe_2$ and VSe_2 (Figures 2c and 2f), and structures ScS_2, $TiTe_2$, and VS_2 (Figures 2b, 2d, and 2e) demonstrated plot overlaps. The accuracy of GPR in its predictions may be attributed to its non-parametric flexibility, enabling the model to capture the intricate interactions and dependencies in the ScS_2 system. This is paired with the ability to quantify certainty due to its Bayesian framework which proves advantageous in dealing with uncertainties in the system's atomic configurations, electronic structures, and other factors. Additionally, GPR is known for its ability to accommodate limited data. These features are the probable cause of the GPR model's higher accuracy compared to other models in this study. Table 2 provides a summary of the prediction accuracy of tested machine learning models.

Table 2. Test Set Prediction Accuracy of ML-Generated MEPs.

	Linear AME	**RFR AME**	**KNN AME**	**GPR AME**
$NbSe_2$	3.48E+02	1.55E-01	1.21E-01	4.46E-02
SCS_2	3.72E+02	1.88E-02	2.01E-02	2.51E-03
$ScSe_2$	1.89E+03	1.25E-02	1.16E-02	5.67E-03
$TiTe_2$	6.67E+02	1.48E-02	5.81E-03	1.09E-04
VS_2	1.07E+03	2.12E-01	8.96E-02	2.28E-03
VSe_2	1.64E+03	1.15E-01	6.77E-02	1.02E-02
All Systems	1684.1028	0.1072	0.0707	0.0093
Training Error	1421.3022	0.0697	0.049	0.0025

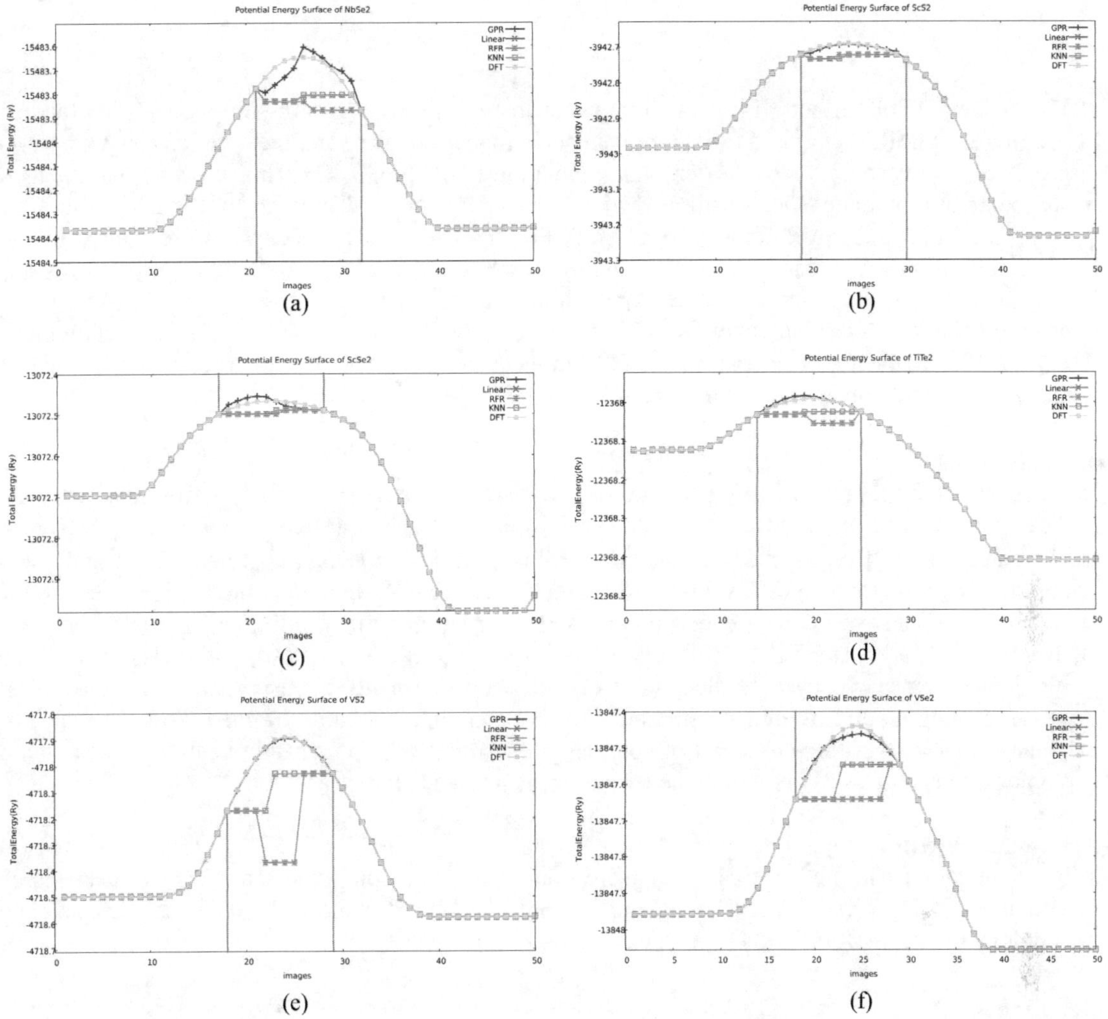

Figure 2. Machine Learning Predictions of the Potential Energy Surface of (a)$NbSe_2$, (b) ScS_2, (c) $ScSe_2$, (d) $TiTe_2$, (e) VS_2, and (f) VSe_2 from the 1H to 1T structural phase.

Table 3. Comparison of ML-Generated to DFT-Calculated Transition Barriers.

	DFT	Linear	Error	RFR	Error	KNN	Error	GPR	Error
$NbSe_2$	0.718	474.7	659.9	0.536	0.254	0.565	0.215	0.764	0.064
SCS_2	0.291	502.6	1728.0	0.260	0.106	0.260	0.106	0.289	0.006
$ScSe_2$	0.230	2967.0	12910.0	0.198	0.137	0.198	0.137	0.242	0.054
$TiTe_2$	0.132	2412.0	18298.0	0.095	0.277	0.102	0.223	0.143	0.088
VS_2	0.608	1413.0	2323.0	0.128	0.790	0.469	0.228	0.593	0.024
VSe_2	0.517	2425.0	4689.0	0.316	0.388	0.411	0.206	0.486	0.060
AME			6768.0		0.325		0.186		0.049

Table 3 shows the DFT-calculated transition barriers in comparison to the ML-generated transition barriers. The Linear Regression model exhibits significantly higher barriers compared to the DFT calculations across all TMDs. This discrepancy confirms that the model is not able to capture the non-linear behavior of the minimum energy pathway curves accurately. The large absolute mean error

(AME) value of 6768 further affirms the limited accuracy of the linear regression model. The Random Forest Regression (RFR) model shows improved performance compared to linear regression, with lower discrepancies between transition barrier values and a lower AME of 0.325. This indicates that the RFR model better approximates the transition barriers, although some deviations from the DFT values are still present. The Gaussian Process Regression (GPR) model exhibits the closest agreement with the DFT-calculated transition barriers, as indicated by its relatively low AME of 0.049. This suggests that the GPR model provides more accurate predictions of the transition barriers for the TMD phase transitions. The K-Nearest Neighbor (KNN) Regression model also performs reasonably well, with an AME of 0.185. While not as accurate as the GPR model, the KNN model shows a closer approximation to the DFT values compared to linear regression and RFR.

4. Conclusions

In this study, the minimum energy pathways and transition barriers of 2D TMDs from the 1H to 1T phase transition were investigated using both density functional theory (DFT) calculations and machine learning algorithms. Results showed that the machine learning models, particularly GPR and KNN Regression, achieved lower AME values, indicating higher accuracy in predicting the minimum energy path- ways compared to Linear Regression and RFR. Furthermore, the runtime comparisons between DFT calculations and machine learning algorithms were conducted to assess their time efficiency. The results indicated that while DFT calculations can be computationally intensive and time-consuming, machine learning models offered a faster and more efficient approach once trained. This advantage of machine learning algorithms in terms of computational time has important implications for the rapid exploration and analysis of minimum energy pathways in TMD phase transitions.

Acknowledgments

The calculations performed in this study were done using the Computing and Archiving Research Environment (COARE) of the Department of Science and Technology-Advanced Science and Technology Institute (DOST-ASTI) in Quezon City, Philippines.

References

[1] Wang Q H *et al* 2012 *Nat. Nanotechnology 7 699*
[2] Wang L S *et al* 2020 *Physical Review B* **102** 2
[3] Mann J *et al* 2014 *Advanced Materials* **26** 9
[4] Zhao B *et al* 2021 *Advanced Functional Materials* **31** 48
[5] Giri A *et al* 2017 *Advanced Materials* **29** 26
[6] Mak K F *et al* 2010 *Physical Review Letters* **105** 136805
[7] Splendiani A *et al* 2010 *Nanoletters* **10** 1271
[8] Qu Y *et al* 2016 Scientific Reports **6** 1
[9] Wu L, *et al.* 2019 *ACS Energy Lett* **4** 7
[10] Butler K T *et al* 2018 *Nature* **559** 7715
[11] Giannozzi P *et al* 2009 Journal of Physics Condensed Matter **21** 39

Journal of Physics: Conference Series **2793** (2024) 012018 doi:10.1088/1742-6596/2793/1/012018

Detection of microplastics in surface water using a DSLR lens-based UV imaging system

RA J Jaruda and JL F Gabayno*

Department of Physics, Mapúa University, Manila, Philippines

⋆ Corresponding author's email: jlfgabayno@mapua.edu.ph

Abstract. Microplastics are any synthetic solid particle or polymeric matrix, ranging from 1 μm to 5 mm in diameter. Due to their low density and size, microplastics can easily be transported through wind and water currents, contaminating the environment with the toxic compounds they contain, harming living organisms through microplastic consumption and exposure. Despite being the top global contributor of plastic waste in 2010, the Philippines still has very limited studies regarding microplastic contamination. There is a need, then, to develop an efficient and accessible means of identifying and quantifying microplastics. To do this, this study utilized a DSLR lens attached to a UV CCD camera to capture images of microplastics, which fluoresce under 395 nm illumination. An image processing algorithm was applied to the resulting images, including binarization, adaptive segmentation, and particle counting to quantify MPs as a fast and easily accessible testing method. Results from the study revealed that despite water samples having been taken within proximity of one another, the number of particles identified within the system still showed a wide spread in values. In addition, it was found that the size distribution of the microplastics obeys the power law. As a new application for UV-based imaging, our method showed that quantitative detection of microplastics is feasible for future in situ measurement of marine water samples. However, further refinement to the counting technique and comparison with standard testing for validation is still recommended for improvement.

1. Introduction

In 2010, the Philippines had been the largest global contributor of plastic waste in the ocean [1]. Various international studies have also determined that microplastics (MPs) have been found to be retained inside the tissues of living things that ingest and cause cellular damage. However, despite the alarming amount of MPs found within the archipelago and its surrounding bodies of water, there is still very limited research done on the extent of MP contamination. Because of this, it is imperative that the degree of MP pollution is determined so as to be aware of the contaminated food and water that is being consumed by humans and animals alike.

MPs are formally defined as any regularly or irregularly shaped synthetic solid particle or polymeric matrix that can range from one micrometer to five millimeters in diameter. They can also be of primary or secondary nature, depending on their manufacturing origin, and are insoluble in water [2].

While the structure and composition of MPs are closely related to the source material and the plastic polymer used, chemical additives are typically added to these plastics for the purpose of production.

These may include toxic compounds such as plasticizers, pigments, antioxidants, acid scavengers, flame retardants, light and heat stabilizers, and lubricants, among others [3, 4].

As a result of their low density and size, MPs can easily be transported through wind and water currents. Additionally, when considering their chemical composition and resistance to biological degradation, MPs easily accumulate and are resistant to environmental decomposition [5]. This can lead to them taking hundreds or even thousands of years to decompose.

Furthermore, the transport of MPs release a mix of hazardous chemicals into ecosystems — a result of the additives added in production to improve the properties of polymers. These toxic chemicals can create negative effects in both the biotic and abiotic environment through possible ingestion, inhalation, or contact [6, 3]. Currently, the factor causing the toxicity of MPs are not yet known — whether it is because of the plastic chemicals or the particles of MPs themselves. Due to their large specific area, however, microplastics are able to interact with toxic compounds such as heavy metals or hold onto organic pollutants, forming more dangerous pollutants [7, 8].

Due to their size, MPs can be especially difficult to detect in surface water as they may easily be mistaken for other particles. As 70% of the time, MP samples are found to be clear [9], it also tends to be difficult to distinguish colorless and amorphous particles of smaller sizes. Hence, an efficient and low-cost imaging system would be beneficial to making MP detection more accessible. With this, the application of ultraviolet (UV) imaging proves to be a promising method for determining the extent of MP pollution as the wavelength dependence of molecular scattering is inversely proportional to the fourth power of the incident wavelength. As such, light scattering off of microplastics is expected to be more significant under UV imaging than under visible light imaging due to the shorter wavelength.

Moreover, MPs polymers have been shown to fluoresce in UV [10], which can enhance the detection sensitivity for these particles. Plastic additives such as bleaching agents are also most often used in textiles and plastic synthesis [11]. The fluorescence induced by these bleaching chemicals under UV would also enhance the imaging system's sensitivity in detecting these MP particles. By positioning the UV CCD camera directly above the MP samples, the reflectance detected from the samples is also optimized. In turn, the aforementioned factors could lead to a sensitive and reliable method for determining the magnitude of MP contamination.

The goal of this study is to develop a UV imaging system that is capable of detecting MPs in surface water. Specifically, this paper aims to construct the UV imaging system by attaching a commercial single lens reflex (SLR) lens to a UV charge-coupled device (CCD) camera under a UV floodlight, identify and quantify the MPs in the water samples, as well as analyze the size distributions of MPs among the different samples.

2. Methodology

2.1. Surface water collection

The surface water samples were collected using a grab sampling method. A 200 mL opaque high-density polyethylene bottle was dipped into the water source and filled with the required sample volume and labeled accordingly with the date and time of collection. As multiple samples were collected within a water source, it was ensured that all samples were collected within a 15-minute time frame. To preserve the sample, the water samples were then directly placed into an ice box to be cooled and were then transferred to a refrigerator to maintain their temperature.

The site for sample collection was on the coast of Manila Bay, along Dolomite Beach, a known location for locals in the capital city of the Philippines. This location was chosen, specifically, as the area has high foot traffic from tourists and locals who frequently visit the artificial beach. The area is also a known fishing site for local fisherfolk and has establishments along its coast such as a private yacht club. In addition, a study by Osorio et al. [12] confirmed that five river mouths of Manila Bay has MP concentrations that ranged from 1,580 to 57,665 particles/m^3 for surface water.

Journal of Physics: Conference Series **2793** (2024) 012018 doi:10.1088/1742-6596/2793/1/012018

Sampling sites along the Dolomite Beach were plotted on a map as shown in figure 1. During the day of sample collection, the weather was recorded to be cloudy and with a surface temperature of 30°*C*.

2.1.1. Descriptions of sample collection sites. Local water hyacinths plants that grow or have washed up along the Dolomite Beach serve as makeshift barriers that intercept plastic trash to reach ashore. These plants were present in all collections sites.

It was observed that collection sites for samples 1, 2, 3, 4, 7, and 9 had no evidently large particles of waste found in the water. However, collection sites for samples 5, 8, and 10 had visible remnants of plastic bags, diapers, and other human-made waste floating on the water.

Increased human activity in the latter collection spots may be considered a factor for the higher concentrations of visible waste.

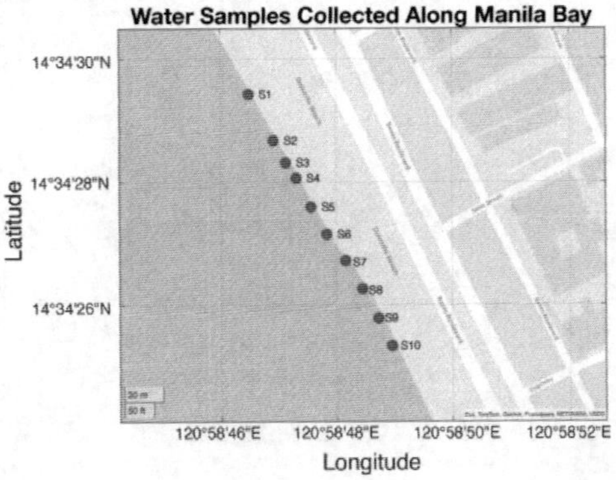

Figure 1. Locations of Collected Samples along Manila Bay, Manila.

2.2. Setup of UV imaging system

The UV imaging system consists of an SLR lens (AF-S DX NIKKOR 18-200mm f/3.5-5.6G ED VR II) was attached to a UV CCD camera (Artray ARTCAM-407UV-WOM) via an F-mount to C-mount lens mount adapter. This was then mounted onto a standard tripod where the lens was set to focus downward in a reflection-type imaging onto the MP samples as shown in figure 2.

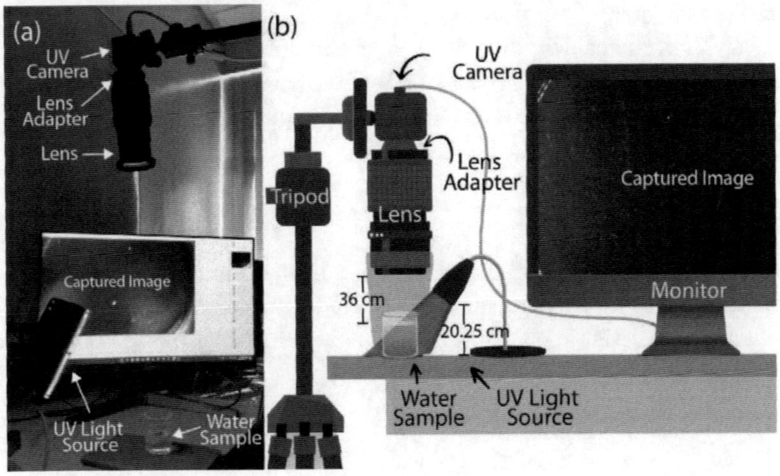

Figure 2. a) UV imaging setup b) Diagram of UV imaging setup.

The samples were then placed onto a beaker to maximize the volume of water samples that can be imaged within the frame that the imaging setup can capture. Moreover, to keep the setup consistent for the imaging of every sample, the distance from the end of the lens was measured to be 36 cm from the surface of the water sample and, the UV floodlight was propped up to a height of 20.25 cm.

2.3. *Resolution of imaging setup*

To determine the resolution of the UV imaging system, the DGK Color Tools High-Resolution Chrome SD Professional Lens Test Chart, as shown in figure 3(a), was used as a basis for the measurement of the real length occupied by one pixel in the image resulting from the imaging setup.

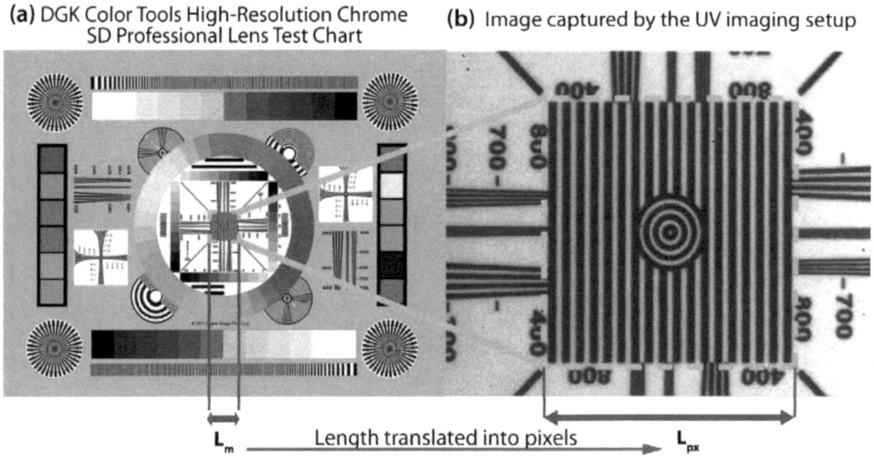

Figure 3. Measurement of the length of a segment in centimeters and in pixels.

$$R = \frac{L_m}{L_{px}} \tag{1}$$

In doing this, equation (1) was used where R refers to the resolution of the imaging system, L_m is the actual length of the segment being measured, and L_{px} is the length of the segment being measured in pixels from the captured image.

As shown in figure 3, the actual length of the segment enclosed in the dashed line box was measured using a caliper and resulted in a measurement of 1.9 cm. Meanwhile, the recorded length of the same segment captured under the UV imaging system spanned 785 px. This resulted to a calculated resolution of 24.2 μm.

$$A = R^2 \tag{2}$$

Moreover, since the particle counting algorithm is set to filter particles by their area, the area of the best resolved pixel was solved for using equation (2). Here, A refers to the area of best resolved pixel while R is the resolution solved for in equation (1). From this, the resulting area of a pixel was calculated to be 576 μm^2.

Journal of Physics: Conference Series **2793** (2024) 012018 doi:10.1088/1742-6596/2793/1/012018

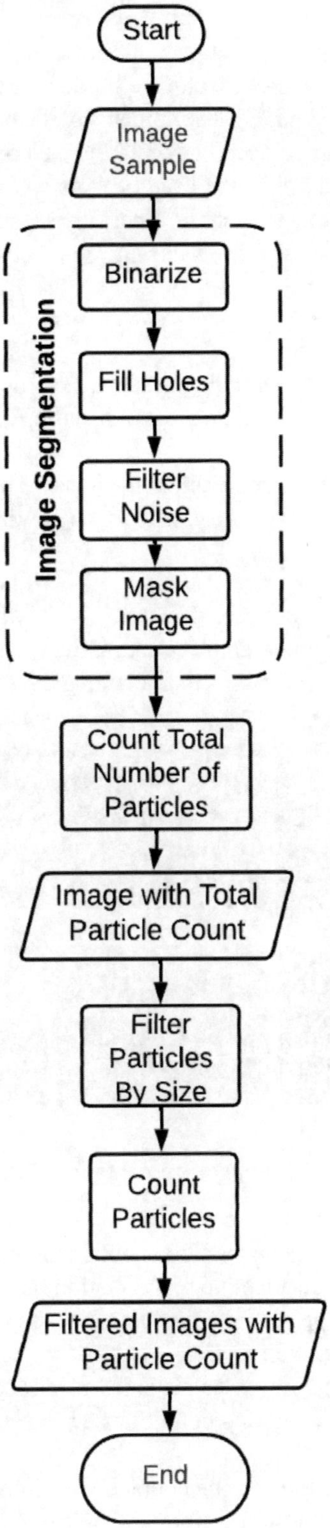

Figure 4. Image Processing Flowchart.

Journal of Physics: Conference Series **2793** (2024) 012018 doi:10.1088/1742-6596/2793/1/012018

2.4. *Water sample preparation*

The ten bottles of surface water samples were taken out of the refrigerator then placed back into an ice box to keep them cool throughout the experiment. One bottle was taken at a time and the surface water was poured onto a 50 mL beaker to be imaged. The contents of the beaker was then poured into a separate HDPE bottle, labelled accordingly, and placed back into the ice box.

In between the imaging of new samples, the used beakers were washed and dried thoroughly to be used for the next batch, ensuring that there were no fibers or particles in the beaker prior to pouring the water sample. This was done by placing the clean and dry beakers under the imaging system and checking for any residual debris.

2.5. *MP detection*

Upon setting up the sample under the UV imaging system, the proprietary Artray imaging software was used to capture the images under the UV light source. All settings in the software were kept consistent across all samples.

Ten images were taken of each water sample under the imaging system so as to maximize the number of particles detected and minimize inconsistencies due to the possible overlap of particles.

2.6. *Image processing*

An overview of the image processing done in this study is shown in figure 4, starting by capturing images of the samples, segmenting them, and the counting the total number of particles in the resulting image. This was followed by filtering the particles in the image by size and counting them.

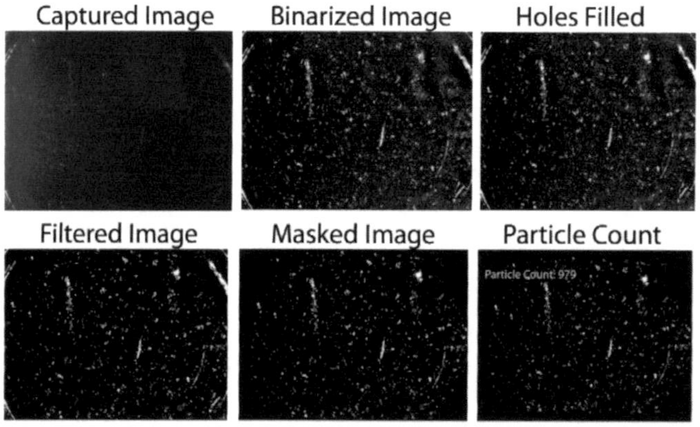

Figure 5. Image Segmentation Process.

2.6.1. Image segmentation. Once all images had been obtained, they were processed using MATLAB. The images were segmented by using an adaptive segmentation function, adaptthresh, as shown in figure 5. This function determines which pixels get thresholder as foreground pixels despite inconsistent lighting within the image. To do this, the function primarily calculates the local mean intensity in the neighborhood of each pixel through first-order statistics to effectively isolate the background from foreground objects throughout varying lighting conditions. In the study, the following syntax was used: T = [adaptthresh, sensitivity].

Here, T is the calculated threshold that can be coupled with imbinarize to convert grayscale images to a binary image. Meanwhile, sensitivity is a scalar, ranging from 0 to 1, which indicates the sensitivity of thresholding more pixels as the foreground. The default value for sensitivity is 0.5 when no value is specified. However, a sensitivity of 0.35 was applied for all raw images as this value effectively separated foreground pixels from the corresponding background pixels with minimal noise. In binarizing the image, the following syntax was used:

Journal of Physics: Conference Series **2793** (2024) 012018 doi:10.1088/1742-6596/2793/1/012018

BW = imbinarize (I,T).

Following this, imfill was used to fill in possible holes in the segmented image that may result from imperfect segmentation. To do this, BW2 = imfill(BW, "holes") was used.

Noise from the resulting image was filtered using bwareafilt. Here, the following syntax was used: BW3 = bwareafilt(BW2, range), where range refers to the minimum and maximum values of an area that needed to be filtered. The range was determined by setting a threshold on the maximum area of the noise — determined by the total area occupied by the biggest fragment — which needed to be removed for a more accurate quantification of particles.

2.6.2. Particle counting. Upon processing the images, the MP particles in the images were identified and quantified through a particle counting algorithm, bwconncomp. This function finds and counts connected components in a binary image with a default connectivity of 8 for 2-D images. Here, the following syntax was used: CC = bwconncomp(BW2).

In addition, particles were sorted in terms of their area as shown in figure 6. This was done by using the bwareafilt function, similar to what was used in the image segmentation process, to remove noise in the images. As the function works by setting the area of pixels to be filtered, the resulting particles were sorted by area instead of diameter as is usually done in related MP studies.

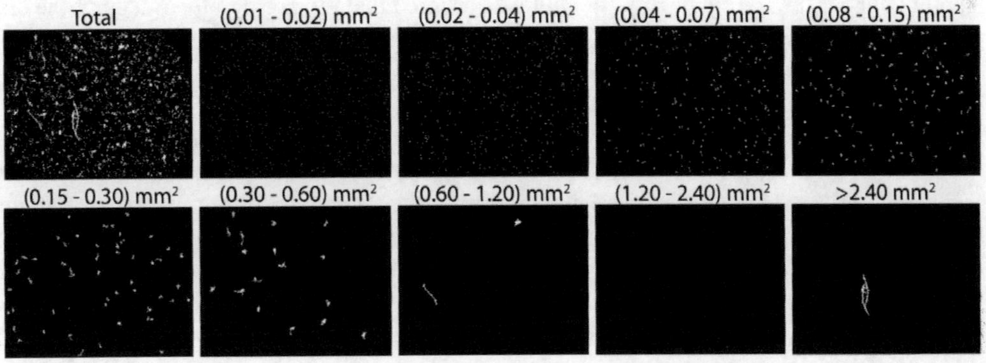

Figure 6. Particle Counting.

Moreover, the filter input filter ranges were determined based on increasing exponents of base 2. As such, table 1 showcases the filter ranges (in pixels) used and their conversion to square millimeters.

It should be noted however that the current system for particle counting has only been verified by visual inspection of the samples as there is no current standard for particle counting among MP samples. A similar method may be tedious to implement on smaller and more irregular particles but could benefit from particle counters or sensors that adhere to calibration standards such as the ISO 2150-4 for particle size and distribution of airborne particles [13]. However, a review on the techniques used for MP quantification have explored the use of microscopy-based, spectroscopy-based, and mass spectrometry-based analyses [11]. Randhawa noted that spectroscopy is frequently used in combination with other methods to count MPs.

Table 1. Conversion of area in pixels to mm^2.

Range	Total area in px	Total area in mm^2
1	16 – 32	0.01 – 0.02
2	33 – 64	0.02 – 0.04
3	65 – 128	0.04 – 0.07
4	129 – 256	0.08 – 0.15
5	257 – 512	0.15 – 0.3
6	513 – 1024	0.3 – 0.6
7	1025 – 2048	0.6 – 1.2
8	2049 – 4096	1.2 – 2.4
9	>4096	> 2.4

In addition, despite visual inspection being the most used and widely available method of identification for MPs [14], it is recommended that chemical characterization such as Fourier transform infrared spectroscopy (FTIR) based methods, and Raman spectroscopy, among others, is undertaken to validate the particles detected by the imaging setup and process.

As for the analysis of the surface water samples, distilled water was used as the control parameter. When captured under the UV imaging system, it was ensured that the noise in the image of the distilled water sample was minimized by applying the bwconncomp function in MATLAB to filter particles in the image that were less than 0.01 mm^2.

3. Results

Figure 7 shows the average particle counts for each sample among ten images taken per sample and the error bars represent the range of the highest and lowest particle counts for a given sample among ten trials. From this, it can be observed that for all samples, there is a decrease in particle count as the size of particles within each given range increases. This is in line with common distributions of MPs as they result from degradation processes.

Meanwhile, the deviations of the particle count among Ranges 1 to 6 can be considered minor. From this, it can be noted that smaller deviations in detected particle counts ensure that the result for those specific samples prove to be more accurate than those with wider spreads in detected particle counts. However, in Ranges 7 and 8, it can be observed that there are instances where one to two particles are detected and instances where there are none for the same sample within the given range. This may be a result of overlapping particles in the image or a simply a result of the movement of particles in and out of the frame of the camera during imaging.

Overall, it is shown that Sample 5 has the highest particle count of all samples with an average of 1749 particles. Sample 5 consistently exhibits the greatest particle count for all ranges with the exception of Range 1 and Range 8. For Range 1, Sample 10 has 139 particles more on average than Sample 5. For Range 8, there were no particles detected for Sample 5. Meanwhile, there was 1 particle detected in Sample 2 and 6, respectively.

The high particle density detected in Sample 5 may be a result of increased human activity around the area as it is most populated by tourists and visitors to the Dolomite Beach. Despite water samples having been taken in proximity of one another, the average particle count of all samples had a spread of 710 at lowest, and 1749 at highest.

Apart from the Control Group, Sample 4 shows the least total particle count with an average of 709. Given this, Sample 4 has the lowest particle count among all ranges except for that of Range 5, 6, and 7.

The differences in the resulting particle counts from the water samples could be a result of differences in fragmentation kinetics within sample locations. Factors affecting this can include environmental elements such as UV radiation, wind, waves, and biodegradation [15]. The density of certain polymer types may also affect the size distributions of particles in the samples as different polymers have varied resistances to fragmentation.

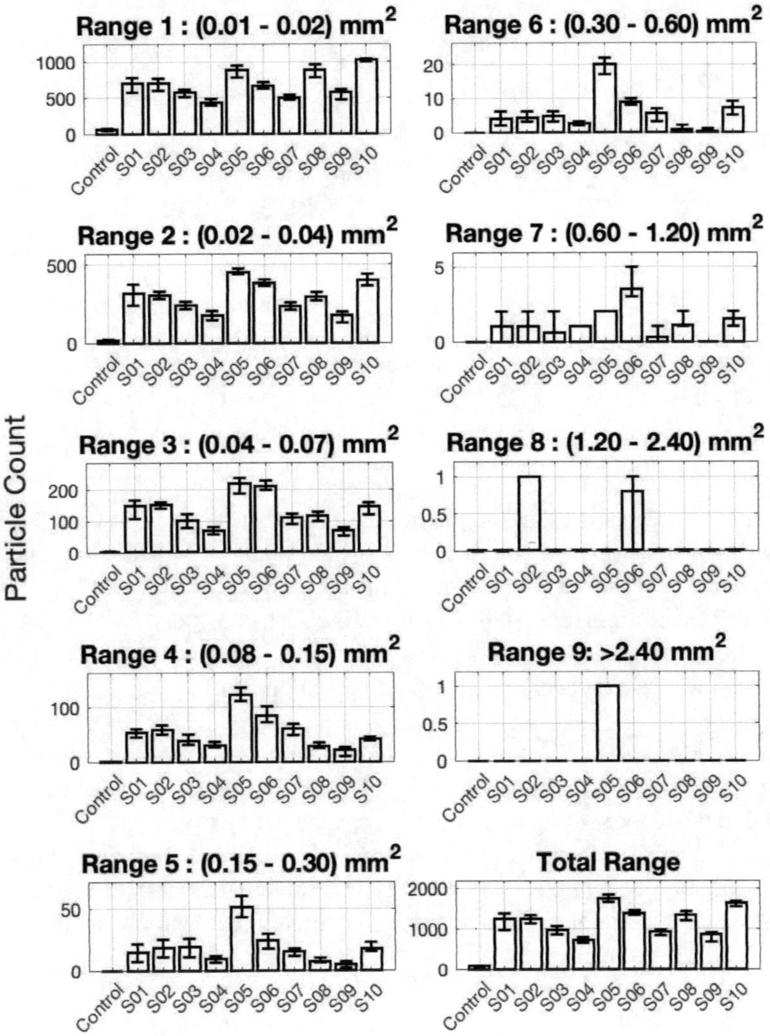

Figure 7. Particle Count vs. Size Distribution for 10 Sampling Locations.

As for the control sample, there was an average of 71.4 particles found overall with no detected particles greater than 0.15 mm^2. The detected particles in the control sample may be regarded as noise as it is significantly less than the number of detected particles the surface water samples.

Meanwhile, figure 8 shows the histogram of Sample 5 on a log-log scale. As the distributions of MPs are known to follow the power law [16, 17], a power law fit was applied to the histograms of water samples in the study. The resulting slopes from the fit ranged from 2.16 to 2.69.

Journal of Physics: Conference Series **2793** (2024) 012018 doi:10.1088/1742-6596/2793/1/012018

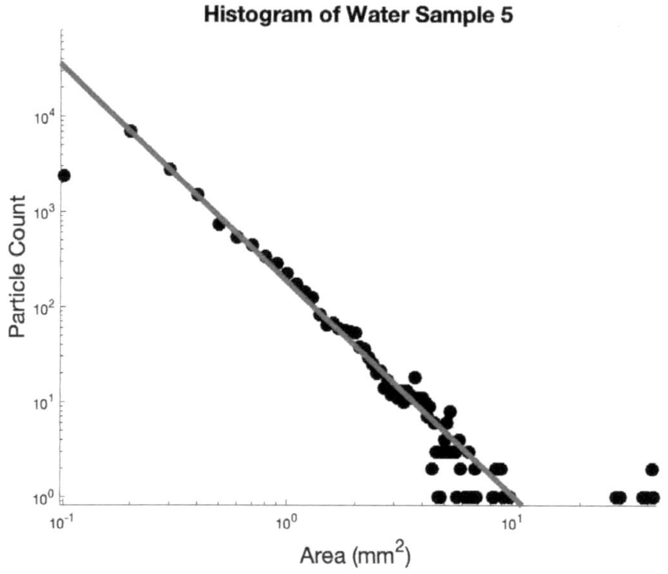

Figure 8. Power Law Distribution of Particle Size with a Slope of 2.25.

For MPs, Kooi et al. [17] noted that the theoretical slope can range from 0 to 3, with a value of 3 representing 3D fragmentation and a value of 1.6 suggesting fragmentation that is partly two-dimensional, as is with plastic sheets. However, their paper from 2021 [16] determined that mean slopes for MP particle area ranged from 1.5 ± 0.0009 for marine surface waters while a value of 2.0 ± 0.065 was obtained for freshwater surface waters. When combining freshwater, marine surface water, and sediments, however, the researchers obtained an average exponent of 2.68.

As the surface water samples collected were taken from along Manila Bay, where waste was highly visible in the water, there was a higher probability of collecting larger particles in the samples taken. This may have led to the higher values of slopes from the power law fit.

Differences in the obtained slopes can also be due to the differences in detection accuracies of the study and of the paper of Kooi et al. [16] where they were able to detect particles as small as 11 or 20 μm. With a lower detection size, the minimum size for which the power law is valid is also lower and therefore may affect the resulting slope when fitted for the power law.

4. Summary and conclusion

In general, the UV imaging setup is capable of detecting particles as small as 0.01 mm^2 (100 μm by 100 μm) and is successful at automated particle quantification in marine water samples. Unlike traditional methods of MP identification such as microscopy, Raman spectroscopy, thermal analysis, and FTIR spectroscopy, the UV imaging system does not require the extraction of MPs from water. As a result, there is no added degradation of MPs in the analyzed samples. There is also no need for the use of chemicals or specialized equipment to treat the samples.

In terms of cost, microscopy techniques are of relatively similar cost with the proposed UV imaging system while thermal analysis as well as Raman and FTIR spectroscopy require expensive and specialized equipment that are not as easily accessible. In terms of terms of accuracy, microscopy is prone to misdiagnoses between synthetic and natural fibers and can take time to count and distinguish MP particles. Raman and FITR spectroscopy, on the other hand, can accurately determine a sample's chemical composition but is time consuming and is unable to discriminate the composition of individual particles. Thermal analysis can accurately determine whether samples are of a plastic material. However, this method degrades samples and, therefore, makes quantification impossible.

In comparison, the UV imaging system is of fairly low-cost as it is composed of commercially available equipment. It is also suitable for rapid and portable detection, being able to detect and quantify particles within minutes after capturing images of the samples. While it is unable to distinguish between organic and synthetic particles, it is a suitable option for preliminary testing of an area for the presence of MPs. By testing an area for MPs using the imaging system, researchers can save on expensive testing fees by isolating water samples with the highest concentration of MP particles prior to sending them to labs. Hence, the imaging system is a fast and accessible way of testing for MP pollution.

It can also be concluded that the use of UV illumination coupled with a UV CCD camera and a commercial DSLR lens can result in the identification of MP particles in surface water samples through image segmentation. From the results, it was also shown that even neighboring samples can greatly differ in MP density and that this can be influenced by the frequency of human activity in a given region. By capturing multiple images of each water sample in succession, the imaging system will also be able to maximize the particles it can identify and quantify.

5. Recommendations

The setup, however, works best for samples with high reflectance under UV, typically polymers. When combined with staining techniques [14], the setup may have increased accuracy in distinguishing between polymers and organic materials. In addition, the wavelength of the UV light source at 395 nm also poses limitations to the reflectance and possible fluorescence that can be observed from the sample. Thus, an improved detection sensitivity of MPs is to be expected with the use of UV-specialized optics or light sources below 395 nm.

Moreover, particles in the sample are still prone to overlapping and may result in less accurate quantification when using the implemented particle counting algorithm. Hence, multiple images were taken of a sample in order to ensure that as the particles move around the medium, they have a chance to separate and be individually counted using the particle counting algorithm. However, a comparison of present results against a particle measuring system which conform to the ISO 2150-4 standard is suggested for future implementation.

Furthermore, it is recommended that the composition of the samples be validated through the use of widely used chemical characterization techniques such as FTIR and Raman spectroscopy.

Acknowledgements

The researchers would like to acknowledge Professor Nobuhiko Sarukura of the Institute of Laser Engineering at Osaka University for the use of the UV CCD Camera and Christian David C. Aquino of the Department of Physics at Mapúa University for his assistance in automating the image processing in MATLAB and help in collecting the water samples the site.

References

[1] Law K L, Starr N, Siegler T R, Jambeck J R, Mallos N J and Leonard G H 2020 *Science Advances* **6**(44)
ISSN 23752548
[2] Frias J P and Nash R 2019 *Marine Pollution Bulletin* **138** 145–147 ISSN 18793363
[3] Campanale C, Massarelli C, Savino I, Locaputo V and Uricchio V F 2020 A detailed review study on potential effects of microplastics and additives of concern on human health
[4] Hahladakis J N, Velis C A, Weber R, Iacovidou E and Purnell P 2018 *Journal of Hazardous Materials* **344** 179–199 ISSN 0304-3894
[5] Syberg K, Khan F R, Selck H, Palmqvist A, Banta G T, Daley J, Sano L and Duhaime M B 2015 *Environmental Toxicology and Chemistry* **34**(5) 945–953 ISSN 1552-8618

[6] Bradney L, Wijesekara H, Palansooriya K N, Obadamudalige N, Bolan N S, Ok Y S, Rinklebe J, Kim K H and Kirkham M B 2019 *Environment International* **131** 104937 ISSN 0160-4120
[7] Andrady A L 2011 Microplastics in the marine environment
[8] Torres F G, Dioses-Salinas D C, Pizarro-Ortega C I and la Torre G E D 2021 *Science of The Total Environment* **757** 143875 ISSN 0048-9697
[9] Löder M G and Gerdts G 2015 *Methodology used for the detection and identification of microplastics—a critical appraisal* (Springer International Publishing) pp 201–227 ISBN 9783319165103
[10] Spizzichino V, Caneve L, Colao F and Ruggiero L 2016 *Applied spectroscopy* **70**
[11] Randhawa J S 2023 *Bulletin of the National Research Centre* **47**(1)
[12] Osorio E D, Tanchuling M A N and Diola M B L D 2021 *Frontiers in Environmental Science* **9** ISSN 2296665X
[13] ISO Central Secretary 2018 Determination of particle size distribution — single particle light interaction methods — part 4: Light scattering airborne particle counter for clean space Standard ISO 21501-4:2018
ISO Geneva, CH
[14] Prata J C, da Costa J P, Duarte A C and Rocha-Santos T 2019 Methods for sampling and detection of microplastics in water and sediment: A critical review
[15] Min K, Cuiffi J D and Mathers R T 2020 *Nature Communications 2020 11:1* **11**(1) 1–11 ISSN 2041-1723
[16] Kooi M, Primpke S, Mintenig S M, Lorenz C, Gerdts G and Koelmans A A 2021 *Water Research* **202** 117429 ISSN 0043-1354
[17] Kooi M and Koelmans A A 2019 *Environmental Science and Technology Letters* **6**(9) 551–557 ISSN 23288930

Journal of Physics: Conference Series **2793** (2024) 012019 doi:10.1088/1742-6596/2793/1/012019

Investigation of a MEMS resonator model with quintic nonlinearity

M. E. Abdelraouf[1,2*], A. Kandil[1,3], W. K. Zahra[1,4] and A. Elsaid [1,5]

[1]Department of Mathematics, Institute of Basic and Applied Sciences, Egypt-Japan University of Science and Technology (E-JUST), New Borg El-Arab City, 21934, Alexandria, Egypt
[2]Department of Basic Sciences, Faculty of Engineering (Shoubra), Benha, 13518, Benha University, Egypt
[3]Department of Physics and Engineering Mathematics, Faculty of Electronic Engineering, Menouf, 32952, Menoufia University, Egypt
[4]Department of Engineering Physics and Mathematics, Faculty of Engineering, Tanta University, Tanta, Egypt
[5]Department of Mathematics and Engineering Physics, Faculty of Engineering, Mansoura University, Mansoura, Egypt

* Corresponding author's email: mohamed.abdelraouf@ejust.edu.eg

Abstract. Micro-electromechanical system (MEMS) resonator is decidedly utilized in a diversity of areas, including time referencing, movement sensing, signal filtration, mass detecting, and further numerous applications. The aim of this article is to use the multiple scales approach to derive analytical formulas for MEMS resonator vibration response. The properties of the complicated nonlinear system at various AC and DC voltages are investigated to be extremely well captured by modeling the dynamics of the micro-beam using multiple scales technique. The resulting Jacobian matrix eigenvalues are tested to verify the stability ranges of these solutions; hence, the jump phenomenon that occurs in experimental performance is interpreted. To study the influence of resonator characteristics on the nonlinear dynamical behavior of such a beam, several response plots are presented. Finally, a numerical solution is obtained with the fourth order Rung-Kutta method to verify the studied model's overall behavior.

1. Introduction

The technology of micro-electromechanical system (MEMS) is an innovative field, and it has significantly evolved the horizons in numerous fields such as healthcare, environmental sensing, industrial automation, Internet of Things (IoT), aerospace, and consumer electronics. MEMS technology enhances functions across these fields by enabling smaller, integrated mechanical and electronic components, thereby increasing efficiency, precision, and versatility compared to non-MEMS alternatives. For healthcare, MEMS sensors were increasingly utilized in wearable and implantable devices to track vital signs, diagnose diseases, and measure the effectiveness of treatment [1, 2]. MEMS sensors were employed in environmental monitoring systems to measure physical environmental parameters such as temperature, pressure, and humidity [3]. In industrial automation systems, MEMS devices provided precise motion control, vibration analysis, and position sensing [4]. In addition, MEMS sensors were integrated into IoT devices for monitoring and control, providing data for predictive

Journal of Physics: Conference Series **2793** (2024) 012019 doi:10.1088/1742-6596/2793/1/012019

maintenance and other applications [5]. MEMS sensors were used in aerospace and defense applications, such as inertial navigation systems, accelerometers, and gyroscopes [6]. Moreover, MEMS technology was used to create small microphones and speakers that can be used to capture and transmit sound in cell phones [7]. In automobiles, MEMS sensors were used to deploy airbags when an accident was on the verge of occurring to protect the occupants [8]. The benefits of MEMS technology extend beyond just the products themselves. MEMS can be used to improve the efficiency of a manufacturing process, leading to lower costs and higher profits. Additionally, MEMS technology can be used to create more precise and accurate measurements, leading to better products and performance. A notable example is MEMS resonators which were immensely used in communication and networking systems to generate precise clocks and timing signals [9].

Over the past two decades, obtaining precise mathematical modelling of nonlinear dynamical systems has had a marked surge of interest. This is because acquiring accurate models of nonlinear dynamical systems results in analysing their complex behaviours, accurately predicting system responses, and discovering efficient control methods. It is widely known that MEMS display a variety of nonlinear behaviors, which have been thoroughly investigated [10]. The ability to optimize, simulate the behavior, and test a system's response without actually building it makes mathematical models crucial since they save costs and aid in creating innovative designs. However, depending on the system being investigated, the difficulty grows as the dynamical system becomes more complicated since nonlinearities resulting from geometry, material, and nonlinear external influences are present. The hardship of dealing with nonlinear equations has heavily spurred a number of academics on studying nonlinear dynamical systems. Perturbation techniques, such as the method of multiple scales and the method of averaging, are powerful tools to study such systems. The most likely reason of this importance is that they provide analytical formulas by which researchers can efficiently investigate the effect of different system's parameters on its response. Nayfeh and Younis [11] utilized the method of multiple scales (MMS) to study a resonant non-linear response that subjected to electric actuation. Piezoelectric, micro electromechanical resonators, that are described as clamped-clamped composite structures, were explored for free and forced oscillations by Li et al. [12].

In this paper, a clamped–clamped MEMS resonator (beam) is introduced to investigate the transversal displacement dominated by quadratic up to quintic nonlinearities. The approximate solution is recovered using the multiple scales approach in order to express the modulations of the amplitude and phase when they are subjected to primary resonance. The Jacobian matrix criterion is employed to provide conditions for stable and unstable solutions. Finally, the entire response of the beam under various operating situations is displayed using numerous curves.

This paper is organized as follows: In section 2, the resonator modeling and physical parameters are introduced. Section 3 explores the perturbation analysis using MMS to the considered resonator. Following this, the results and graphs are introduced and discussed. The last section offers a conclusion of the study.

2. MEMS resonator modelling

'Figure 1' shows a schematic diagram for the clamped-clamped micro-beam under the effect of direct current (DC) source and alternating current (AC) source with frequency Ω. The physical parameters of the resonator are: Young modulus $E = 16 * 10^{10}$ Pa, mass density $\rho = 2.5 * 10^3$ Kg/m^3, length $L = 25 * 10^{-5}\ m$, width $b = 5 * 10^{-5}\ m$, and depth $h = 2.7 * 10^{-6}\ m$. The resonator is one moving electrode which is separated by a gap distance $d = 2.5 * 10^{-6}\ m$ from a lower stationary electrode [13]. The resonator is considered as a Euler-Bernoulli beam; therefore, the equation that describes the vertical displacement $\widetilde{w}$ of the resonator through the length $\tilde{x}$ and the time $\tilde{t}$ is given by

$$EI\frac{\partial^4\widetilde{w}}{\partial\tilde{x}^4} + m\frac{\partial^2\widetilde{w}}{\partial\tilde{t}^2} + c\frac{\partial\widetilde{w}}{\partial\tilde{t}} - \widetilde{N}\frac{\partial^2\widetilde{w}}{\partial\tilde{x}^2} = \frac{EA}{2L}\left(\int_0^L\left(\frac{\partial\widetilde{w}}{\partial\tilde{x}}\right)^2 d\tilde{x}\right)\frac{\partial^2\widetilde{w}}{\partial\tilde{x}^2} + \frac{b\,\epsilon\,(V_{ac}\cos(\Omega\,\tilde{t}) + V_{dc})^2}{2(d-\widetilde{w})^2}, \quad (1)$$

where $I = b\,h^3/12$ is the area moment of inertia, m is the beam's mass per unit length, c is the viscous damping coefficient, ϵ is the dielectric constant, and A is the cross-sectional area of the beam. The beam is subjected to the fixed-fixed boundary conditions. Given that ~ identifies the dimensional quantities and those without ~ are dimensionless, Eq. (1) is nondimensionalized to be

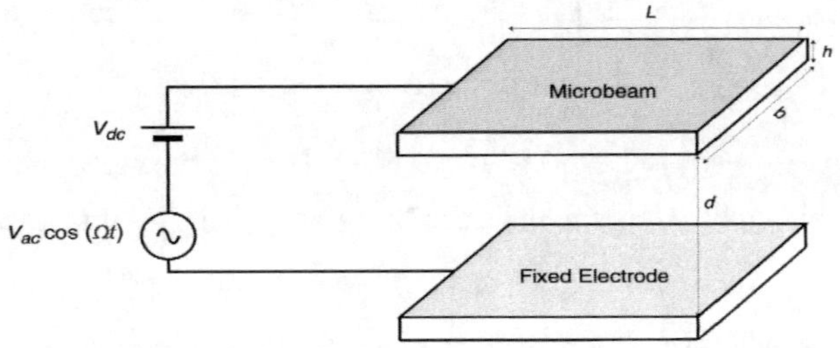

Figure 1. The MEMS resonator's schematic diagram.

$$\frac{\partial^4 w}{\partial x^4} + \frac{\partial^2 w}{\partial t^2} + 2\mu \frac{\partial w}{\partial t} - N\frac{\partial^2 w}{\partial x^2} - \alpha_1 \left(\int_0^1 \left(\frac{\partial w}{\partial x}\right)^2 dx\right)\frac{\partial^2 w}{\partial x^2} = \frac{\beta\,(V_{ac}\cos(\Omega\, t) + V_{dc})^2}{2(1-w)^2}, \quad (2)$$

where $\tilde{w} = w\,d$, $\tilde{x} = x\,L$, $t = \tilde{t}\sqrt{\frac{EI}{m\,L^4}}$, $2\mu = c\sqrt{\frac{L^4}{EIm}}$, $N = \tilde{N}\frac{L^2}{EI}$, $\alpha_1 = \frac{6\,d^2}{h^2}$, and $\beta = \frac{b\,\epsilon\,L^4}{2EId^3}$. The solution of Eq.(2) is the sum of static and dynamic deflections. The static deflection w_s is considered when the change of transverse motion with time is zero,

$$\left(\frac{\partial^4 w_s}{\partial x^4} - N\frac{\partial^2 w_s}{\partial x^2} - \alpha_1\left(\int_0^1 \left(\frac{\partial w_s}{\partial x}\right)^2 dx\right)\frac{\partial^2 w_s}{\partial x^2}\right)(1-w_s)^2 = \beta\left(\frac{V_{ac}^2}{2} + V_{dc}^2\right). \quad (3)$$

The static deflection w_s is estimated numerically to be $w_s = 0.0262\,x^2\,(x-1)^2$.
Based on primary resonance case, a single mode dynamic deflection is assumed as $w_d(x,t) = \phi(x)\,u(t)$, where $\phi(x)$ is the spatial function has the form

$$\phi(x) = 0.982\,sin(4.596\,x) - cos(4.596\,x) + cosh(4.596\,x) - 0.982\,sinh(4.596\,x), \quad (4)$$

and $u(t)$ is the temporal function and it is obtained by solving the following the equation

$$\ddot{u} + 2\mu\,\dot{u} + \omega^2\,u + \delta_1 u\ddot{u} + \delta_2 u^2\ddot{u} + \alpha_{23}u\dot{u} + \alpha_{33}u^2\dot{u} + \alpha_{21}u^2 + \alpha_{31}u^3 + \alpha_4 u^4 + \alpha_5 u^5 = k_1\,cos(\Omega\,t) + k_2\,cos(2\Omega\,t), \quad (5)$$

where the coefficients of Eq. (5) are carefully obtained from [13].

3. Perturbation analysis

Using MMS to attack Eq. (5), a semi-closed form of Eq. (5) is assumed in the form

$$u(t\,;\,\varepsilon) = u_0(T_0,T_1,T_2) + \varepsilon\,u_1(T_0,T_1,T_2) + \varepsilon^2\,u_2(T_0,T_1,T_2) + \cdots, \quad (6)$$

where $T_0 = t$, $T_1 = \varepsilon\,t$, and $T_2 = \varepsilon^2\,t$ are the time scales and ε is a perturbation parameter. The perturbed Eq. (5) takes the form

$$\ddot{u} + 2\varepsilon^2\,\mu\,\dot{u} + \omega^2\,u + \varepsilon\,\delta_1 u\ddot{u} + \varepsilon^2\delta_2 u^2\ddot{u} + \varepsilon\,\alpha_{23}u\dot{u} + \varepsilon^2\alpha_{33}u^2\dot{u} + \varepsilon\,\alpha_{21}u^2 + \varepsilon^2\alpha_{31}u^3 + \varepsilon\alpha_4 u^4 + \varepsilon^2\alpha_5 u^5 = \varepsilon^2\,k_1\,cos(\Omega\,t) + \varepsilon^2\,k_2\,cos(2\Omega\,t), \quad (7)$$

Using the chain rule, $\dot{u}$ and $\ddot{u}$ take the forms $\dot{u} = (D_0 + \varepsilon\,D_1 + \varepsilon^2\,D_2)u$, $\ddot{u} = (D_0 + \varepsilon\,D_1 + \varepsilon^2\,D_2)^2\,u$ where D_0, D_1, and D_2 are the partial differential operators with respect to T_0, T_1, and T_2, respectively. The primary resonance case is considered, $\Omega = \omega + \varepsilon^2\sigma$, where σ is the detuning parameter which measures the nearness of the excitation frequency from the natural frequency. Performing the substitution into Eq. (7) and solving the different scales equations with avoiding unbounded solutions yield the frequency response equation. This equation describes the effect of the change in excitation frequency 'σ' and its intensity 'k_1' on the vibration amplitude 'a' as follows

$$\left(\sigma\,\mathrm{a} + \frac{\mathrm{a}^3}{\omega}\left(\frac{\alpha_{23}^2}{24} - \frac{3\alpha_{31}}{8} - \frac{11\alpha_{21}\delta_1}{24} + \frac{5\alpha_{21}^2}{12\omega^2} + \frac{\delta_1^2\omega^2}{24} + \frac{3\delta_2\,\omega^2}{28}\right) + \frac{\mathrm{a}^5}{\omega}\left(-\frac{5\alpha_5}{16} - \frac{9\alpha_4\delta_1}{16} + \frac{7\alpha_4\alpha_{21}}{8\omega^2}\right) + \frac{63\,\alpha_4^2}{160\omega^3}\mathrm{a}^7\right)^2 + \left(\mu\,\mathrm{a} + \frac{\mathrm{a}^3}{8}\left(\alpha_{33} - \frac{\alpha_{21}\alpha_{23}}{\omega^2}\right) - \frac{\alpha_4\alpha_{23}}{16\omega^2}\mathrm{a}^5\right)^2 - \frac{k_1^2}{4\omega^2} = 0. \quad (8)$$

In addition to the frequency response equation, the stability regions are deduced by checking the eigenvalues of the Jacobian matrix $J = \begin{bmatrix} A & B \\ C & D \end{bmatrix}$, where the elements of J are

$$A = -\mu - \frac{3\mathrm{a}^2}{8}\left(\alpha_{33} - \frac{\alpha_{21}\alpha_{23}}{\omega^2}\right) + \frac{5\alpha_4\alpha_{23}}{16\omega^2}\mathrm{a}^4, \quad (9)$$

$$B = -\sigma\, \mathrm{a} - \frac{m\, \mathrm{a}^3}{\omega} - \frac{\mathrm{a}^5}{\omega}\left(-\frac{5\alpha_5}{16} - \frac{9\alpha_4\delta_1}{16} + \frac{7\alpha_4\alpha_{21}}{8\omega^2}\right) + \frac{63\, \alpha_4^2}{160\omega^3}\mathrm{a}^7, \tag{10}$$

$$C = \frac{\sigma}{\mathrm{a}} + \frac{3z\mathrm{a}}{\omega} + \frac{5\mathrm{a}^3}{\omega}\left(-\frac{5\alpha_5}{16} - \frac{9\alpha_4\delta_1}{16} + \frac{7\alpha_4\alpha_{21}}{8\omega^2}\right) + \frac{441\, \alpha_4^2}{160\omega^3}\mathrm{a}^5, \tag{11}$$

$$D = -\mu - \frac{\mathrm{a}^2}{8}\left(\alpha_{33} - \frac{\alpha_{21}\alpha_{23}}{\omega^2}\right) + \frac{\alpha_4\alpha_{23}}{16\omega^2}\mathrm{a}^4, \tag{12}$$

where $z = \left(\frac{\alpha_{23}^2}{24} - \frac{3\alpha_{31}}{8} - \frac{11\alpha_{21}\delta_1}{24} + \frac{5\alpha_{21}^2}{12\omega^2} + \frac{\delta_1^2\omega^2}{24} + \frac{3\delta_2\,\omega^2}{28}\right)$. Hence, the steady-state solution is stable if the real parts of the Jacobian matrix eigenvalues are negative, otherwise it is unstable.

4. Results and discussion

The response of the MEMS resonator is examined by adjusting the frequency of the AC supply close to the resonator's natural frequency, $\omega = 21.1253$. In the upcoming figures, the solid line denotes stable response and the dashed one represents unstable response. Parametric variations, including Ω and $\boldsymbol{V_{dc}}$ are carried out to see the effect it would have on the system behaviour.

'Figure 2' compares frequency response curves for different values of $\boldsymbol{V_{dc}}$ and keeping $\boldsymbol{V_{ac} = 10}$. It shows a linear response in which by increasing the detuning parameter σ, the amplitude 'a' increases until it reaches the peak value at $\sigma = 0.057$. Moreover, despite the rise of the frequency, the amplitude value decreases regularly. The resonator exhibits a collateral hardening behaviour to the jump phenomena which begins to appear with increasing $\boldsymbol{V_{dc}}$ value as shown in Figs. 2(B) – 2(D). In Figs. 2(B) and 2(C), not only the jump phenomena and consequently the nonlinear response can be clearly noted when the value of the $\boldsymbol{V_{dc}}$ goes higher, but the peak amplitude also increases significantly. In the case of Fig. 2(D), when the frequency of the AC source Ω changes from small values to higher ones (forward sweeping), the response amplitude is observed to soar rapidly till the saddle-node bifurcation point A which corresponds to $\sigma = 1.26$. At this point, a sudden jump occurs to point B after which the amplitude decays smoothly. In addition, when σ is swept down, what emerges is that the amplitude, 'a', increases gradually until the saddle-node bifurcation point C which corresponds to $\sigma = 0.32$. In a further drop in σ, a remarkable transition occurs to point D, after which the amplitude, 'a', decreases steadily.

'Figure 3' and 'Figure 4' illustrates the dynamical response of the MEMS resonator during forward sweeping for $\boldsymbol{V_{dc} = 10}$ and $\boldsymbol{V_{ac} = 10}$ at $\sigma = 0.67$ and $\sigma = 0.68$, respectively. Not only do these two figures depict unambiguously the jump phenomena represented by sudden reduction of the amplitude from 0.2017 to 0.0243, but they also illustrate a stable limit cycle.

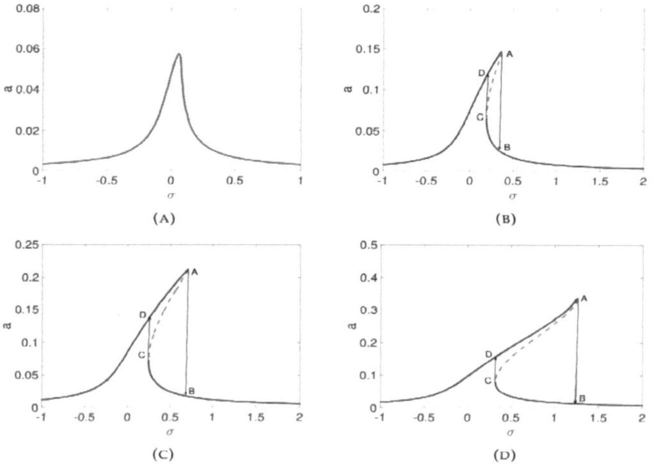

Figure 2. Frequency response curves of the resonator at $\boldsymbol{V_{ac} = 10}$ with (A) $\boldsymbol{V_{dc} = 2}$ (B) $\boldsymbol{V_{dc} = 5}$ (C) $\boldsymbol{V_{dc} = 7}$ (D) $\boldsymbol{V_{dc} = 10}$

'Figure 5' describes the time history and phase plane with backward sweeping at $\sigma = 0.42, 0.41$. It can be observed that the jump occurs in the amplitude unexpectedly from the value 0.0415 to 0.1669.

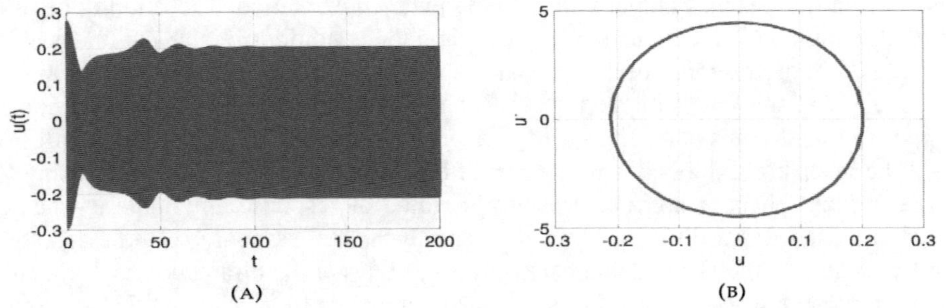

Figure 3. Dynamical response of the MEMS resonator during forward sweeping at $\sigma = 0.67$ for $\boldsymbol{V_{dc}} = \mathbf{10}$ and $\boldsymbol{V_{ac}} = \mathbf{10}$ (A) Time history (B) Phase plane

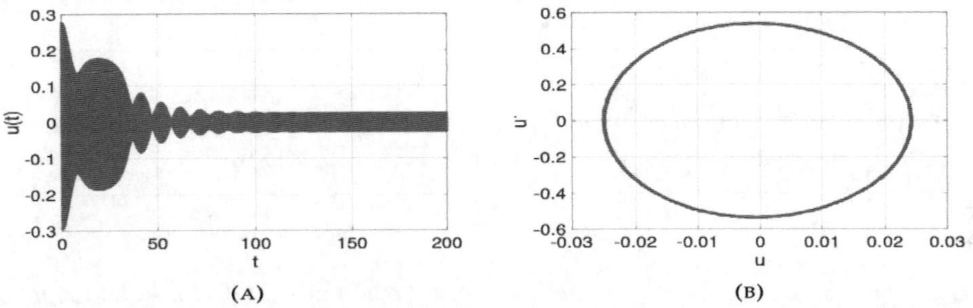

Figure 4. Dynamical response of the MEMS resonator during forward sweeping at $\sigma = 0.68$ for $\boldsymbol{V_{dc}} = \mathbf{10}$ and $\boldsymbol{V_{ac}} = \mathbf{10}$ (A) Time history (B) Phase plane

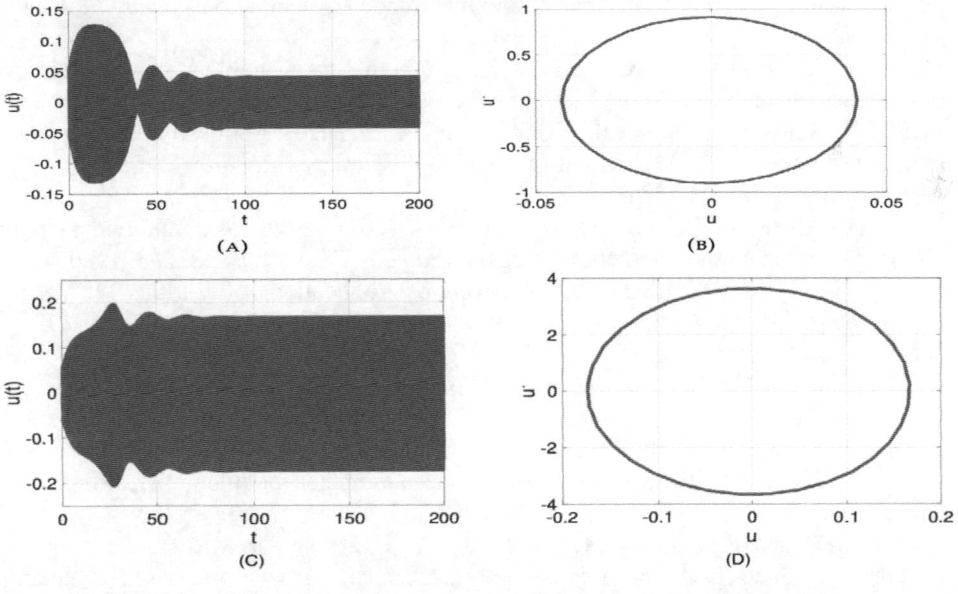

Figure 5. Dynamical response of the MEMS resonator during backward sweeping for $\boldsymbol{V_{dc}} = \mathbf{10}$ and $\boldsymbol{V_{ac}} = \mathbf{10}$ (A) Time history at $\sigma = 0.42$ (B) Phase plane at $\sigma = 0.42$ (C) Time history at $\sigma = 0.41$ (D) Phase plane at $\sigma = 0.41$

5. Conclusion

This study focuses on investigating a clamped-clamped MEMS resonator to examine the vertical displacement in the presence of quadratic up to quintic nonlinearities. The resonator was under the effect of electric AC and DC supplies and accounts for the axial load and mid-plane stretching. The method of multiple scales was adopted to obtain approximate solutions and analytical formulas for determining the tonality of response amplitude and phase under the effect of primary resonance, indicating the degree of distinct patterns in the system's output with varying external excitation. The stability of solutions was scrutinized depending on Jacobian matrix analysis, and the conditions of stable and unstable solutions were attained. The entire response of the resonator under various operating situations was displayed, in addition to numerical analysis validation using Runge-Kutta method. In frequency-response curves, a linear response was noted at small values of $\boldsymbol{V_{dc}}$, and a nonlinear scheme started to manifest with rising $\boldsymbol{V_{dc}}$ value. The resonator exhibited a hardening effect, which expanded with increasing $\boldsymbol{V_{dc}}$ values. Furthermore, a jump phenomenon occurred in the value of response amplitude a when the supply frequency Ω increased forwardly or backwardly. The dynamical response of the resonator showed a stable limit cycle during the forward and backward sweepings at the bifurcation saddle-node points on frequency-response curves. This study introduces a novel insight by elucidating that the observed jump phenomena in experiments result from the presence of unstable regions where the resonator ceases to function. Analytical determination of these unstable conditions is accomplished through the application of the multiple scales method. Future research in controlling system behaviour offers the potential to improve operational efficiency and performance.

Acknowledgements

This research work is supported by the Egyptian Ministry of Higher Education (MOHE) grant in the scope of the Egypt-Japan University of Science and Technology(E-JUST).

References

[1] Basu, A. K., Basu, A., Ghosh, S., & Bhattacharya, S. (Eds.). (2021). *MEMS Applications in Biology and Healthcare*. Melville: AIP Publishing.

[2] Chen, Z., Kan, X., Yuan, Q., Wang, T., Yang, J., & Yang, F. (2020). A switchable high-performance RF-MEMS resonator with flexible frequency generations. *Scientific Reports*, *10*(1), 4795.

[3] Cureau, R. J., Pigliautile, I., & Pisello, A. L. (2022). A new wearable system for sensing outdoor environmental conditions for monitoring hyper-microclimate. *Sensors*, *22*(2), 502.

[4] Faudzi, A. A. M., Sabzehmeidani, Y., & Suzumori, K. (2020). Application of micro-electro-mechanical systems (MEMS) as sensors: a review. *Journal of Robotics and Mechatronics*, *32*(2), 281-288.

[5] Coulby, G., Clear, A. K., Jones, O., & Godfrey, A. (2021). Low-cost, multimodal environmental monitoring based on the Internet of Things. *Building and Environment*, *203*, 108014.

[6] Ulla, S. Z., & Vura, S. (2021). An Analysis of Applications of Various MEMS Sensors in Aerospace Industry. *Journal of Aerospace Engineering & Technology*, *11*(1), 16-20.

[7] Tiete, J., Domínguez, F., da Silva, B., Touhafi, A., & Steenhaut, K. (2017). MEMS microphones for wireless applications. In *Wireless MEMS Networks and Applications* (pp. 177-195). Woodhead Publishing.

[8] Ramakrishnan, J., Gaurav, P. R., Chandar, N. S., & Sudharsan, N. M. (2021). Structural design, analysis and DOE of MEMS-based capacitive accelerometer for automotive airbag application. *Microsystem Technologies*, *27*, 763-777.

[9] Zhao, W., Ahmed, S., Fariborzi, H., & Younis, M. I. (2021). Analytical and experimental study of the dynamics of a micro-electromechanical resonator based digital-to-analog converter. *Journal of Micromechanics and Microengineering*, *31*(12), 125010.

[10] Hajjaj, A. Z., Jaber, N., Ilyas, S., Alfosail, F. K., & Younis, M. I. (2020). Linear and nonlinear dynamics of micro and nano-resonators: Review of recent advances. *International Journal of Non-Linear Mechanics*, *119*, 103328.

[11] Younis, M. I., & Nayfeh, A. H. (2003). A study of the nonlinear response of a resonant microbeam to an electric actuation. *Nonlinear dynamics*, *31*, 91-117.

[12] Li, H., & Balachandran, B. (2006). Buckling and free oscillations of composite microresonators. *Journal of Microelectromechanical Systems*, *15*(1), 42-51.

[13] Rocha, R. T., Alfosail, F., Zhao, W., Younis, M. I., & Masri, S. F. (2021). Nonparametric identification of a micro-electromechanical resonator. *Mechanical Systems and Signal Processing*, *161*, 107932.

A Monte Carlo study of K escape phenomena of X-ray spectra in CdTe and Gd_2O_2S detectors

Dan Wang[1,2], Zhe Shi[1,2] and Zhi Yang[1,2,*]

[1]School of Biomedical Engineering, Capital Medical University, Beijing, China;
[2]Beijing Key Laboratory of Fundamental Research on Biomechanics in Clinical Application, Capital Medical University, Beijing, China

* Corresponding author's email: zhiyang@ccmu.edu.cn

Abstract. Detectors can be categorized as energy-integrated detectors (EID) and photon-counting detectors (PCD). Both of them show K x-ray escape effects, leading to potential inaccuracies in spectral data. This paper focuses on investigation of the detector's K x-ray escape effects on spectra. Geant4-based Monte Carlo simulations were initially validated against literature results. Subsequently, the X-ray escape events and spectra in both a CdTe (PCD) and Gd_2O_2S (EID) were explored. X-ray spectra generated using Geant4 demonstrated close agreement with the existing literature. The spectra with typical absorption edges were observed in both detectors. This study highlights the successful application of Monte Carlo simulations for spectral measurements and emphasizes impact of K escape effects on X-ray spectra. The positions of escape events are contingent upon the element properties in the detectors, providing valuable insights into understanding K escape effects.

1. Introduction

X-ray detectors encompass two categories: photon-counting detectors (PCD) and integrating detectors (EID) [1-4]. The detector serves as a crucial component in computed tomography (CT), influencing directly on image quality [5-7]. PCDs leverage semiconductors to transform X-ray photons into electron-hole pairs. Upon the application of voltage across the semiconductor, electrons migrate toward the anode, where they are collected, thereby generating a signal. In EIDs, scintillators facilitate the conversion of X-ray photons into visible light [8-9].

The performance assessment metrics for detectors include detection efficiency, resolution, stopping power, and other factors. The escape of X photons can result in a reduction of detection efficiency and introduce inaccuracies in spectra. Nonetheless, achieving precise X-ray spectrum measurements remains essential for CT imaging systems. Escape may be caused by the characteristic X-ray escape, Compton scattering , and photoelectron escape, in detectors [10]. Photons with energy just above K-edge undergo photoelectric interactions. When the atoms transits to the ground level they often emit a characteristic X-ray. Concurrently, these X-rays may exit the detector volume, resulting in only a fraction of the photon energy being deposited. These occurrences contribute to increased occurrences of the low-energy events, commonly denoted as escape effects [11]. Escape fraction is strongly related to detector materials and thickness. Besides, photoelectron escape is analogous to K-escape, predominantly occurring in thinner detector configurations [12].

Journal of Physics: Conference Series **2793** (2024) 012020 doi:10.1088/1742-6596/2793/1/012020

When measuring quantitative spectra, the K-edge of detector materials serves as a pivotal criterion for detector selection. The escape phenomenon will vary depending on the detector material characteristics. If the K-edge of the detector is higher than energy of the incident photon, then correction for the detector's escape effect is unnecessary.

Analysis of escape phenomenon in spectra for detectors cannot be generalized. Different detectors should be investigated independently. Popular detectors include Gd_2O_2S [12-13], Si [14], CdTe [15-17], CsI [18], etc. CdTe is an optimal choice for quantitative measure of X-ray spectroscopy owing to its significantly higher stopping power. Additionally, it is a candidate for PCDs. In terms of CdTe, a developed commercial software has been implemented for X-ray spectrum quantitative correction [17]. Monte Carlo method has been employed for fitting the escape fraction and it shows that escape peaks have a pronounced effect in CdTe slab [19]. What's more, detector response function were calculated with the Monte Carlo method for the escape correction of X-ray spectra of CdTe, and the final escape correction results were compared with the measurement results of high-purity germanium [16,19]. Further, obtaining the distribution angle of photoelectrons, the photoelectron escape fraction can be calculated to achieve photoelectron escape correction of the measured spectra [20-21]. In addition, Gd_2O_2S is one of the most popular EIDs. Although Gd_2O_2S material is not the candidate for X-ray spectroscopy measurement, observation for escape phenomena of spectra on this detector will be also useful. Studies have shown that the escape efficiency [22] of incident photons reaches 31.3% at 55 keV (slightly higher than the K-edge of Gd), when the detector's thickness is 90 mg/cm^2 [12]. In detector model, the escape effect usually is taken into account [23].

Although escape efficiency for Gd_2O_2S has been investigated, the material has not yet been employed for spectral measurement. This study will use Geant4-based Monte Carlo programs to achieve X-ray spectra measurement on CdTe and Gd_2O_2S materials. Distinct escape phenomena of two type detectors will be examined.

2. Materials and methods

2.1. Monte Carlo

Geant4 is a Monte Carlo simulation software implemented in C++. Almost all aspects of the physical simulation process have been included in the software. It offers developer opportunities to construct specific modules by themselves. The data were recorded automatically using the data analysis framework (ROOT). More details and guidelines for data analysis can be found in the official Handbook of Geant4.

2.2. Geometry

Figure 1 (a) represents a full description of the X-ray tube geometry used in this work. The anode and the source of electrons were shown (a parallel electron beam of size 1 mm) [24]. The X-rays emitted from the anode pass through the filtration. Figure 1 (b) shows the reconstructed model by Geant4. Red lines are the trajectories for electrons.

Journal of Physics: Conference Series **2793** (2024) 012020 doi:10.1088/1742-6596/2793/1/012020

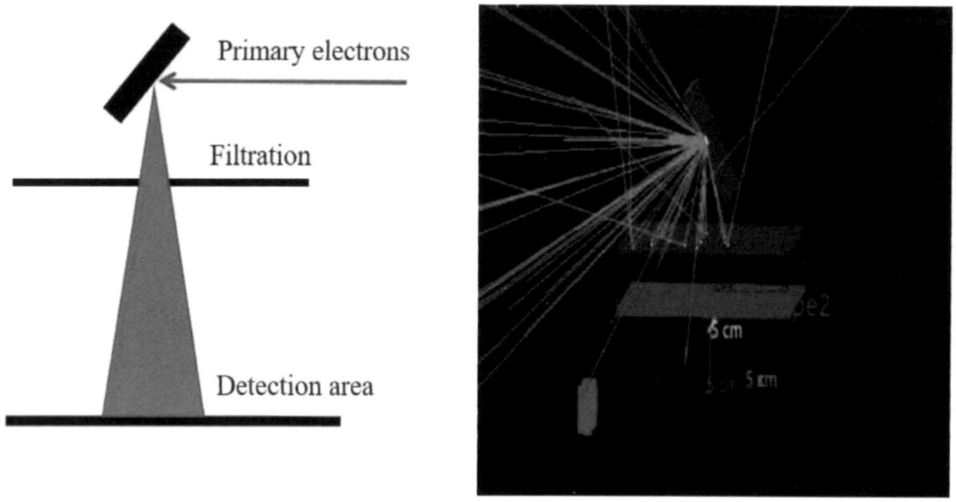

(a) (b)

Figure 1. (a) Description of the simulated X-ray tube model. (b) Geant4 model, red lines are the trajectories for electrons.

3. Results

3.1. Validation of simulated X-ray spectra against experimental data

Simulated X-ray spectra and the actual spectra obtained from reference [25] were compared. GE Maxiray (125B) X-ray tube was used in the study. Its parameters are 2.8 mm Al filtration, tungsten anode with 10° fixed angle, and a detector 1 m away from the source. we compared X-ray spectra at tube voltages of 70 kVp and 140 kVp. 10^8 x-ray electrons were input. The spectra are recorded by data analysis framework (ROOT) supported by Geant4. Experimental and simulated results match well as shown in figure 2. It implies the measurement method and the code are reliable.

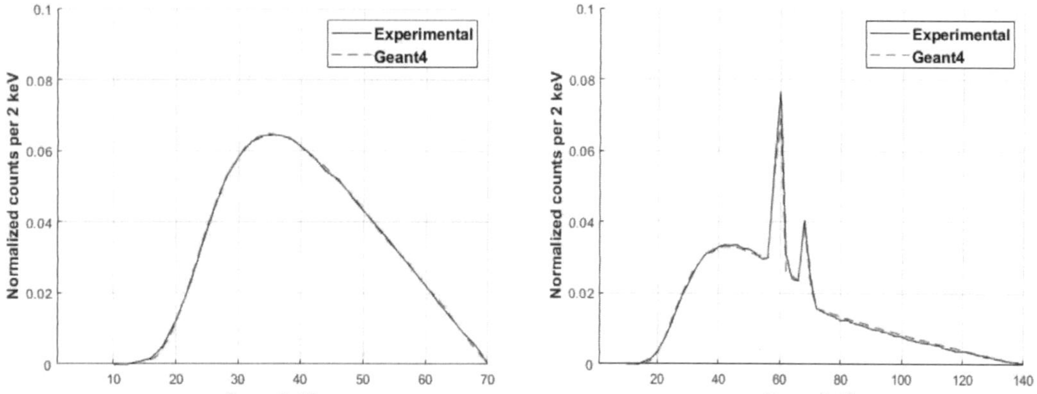

(a) (b)

Figure 2. Comparison of simulated X-ray spectra with experimental literature data [25]. Results for two tube voltages, (a)70 kVp and (b) 140 kVp.

3.2. Measurement of spectrum by CdTe

X-ray spectrum was measured by an CdTe detector. The thickness is 1 mm. The tube voltage was set to 80 kVp with a tungsten anode and a 1.3 mm aluminum filter. The anode angle was fixed at 7° and the direction of electron beam is shown in figure 3(a). 2×10^9 electrons were injected. The flowchart for measuring escape events is displayed in figure 3(b). If a photon leaves the detector and partial energy is deposited in detector at the same time, it is considered as the escape event. The effect of scattering and detector sizes are not considered in this study.

Discontinuities in the original spectrum (figure 3(c), black line) were observed at 27 and 32 keV due to the escape of the K-fluorescent X-rays of Cd and Te. Although the filtration should stop essentially all the incident photons on the detector at low energy, many lower energy events were observed (green line). The comparable phenomenons were revealed as well in published papers [16-17].

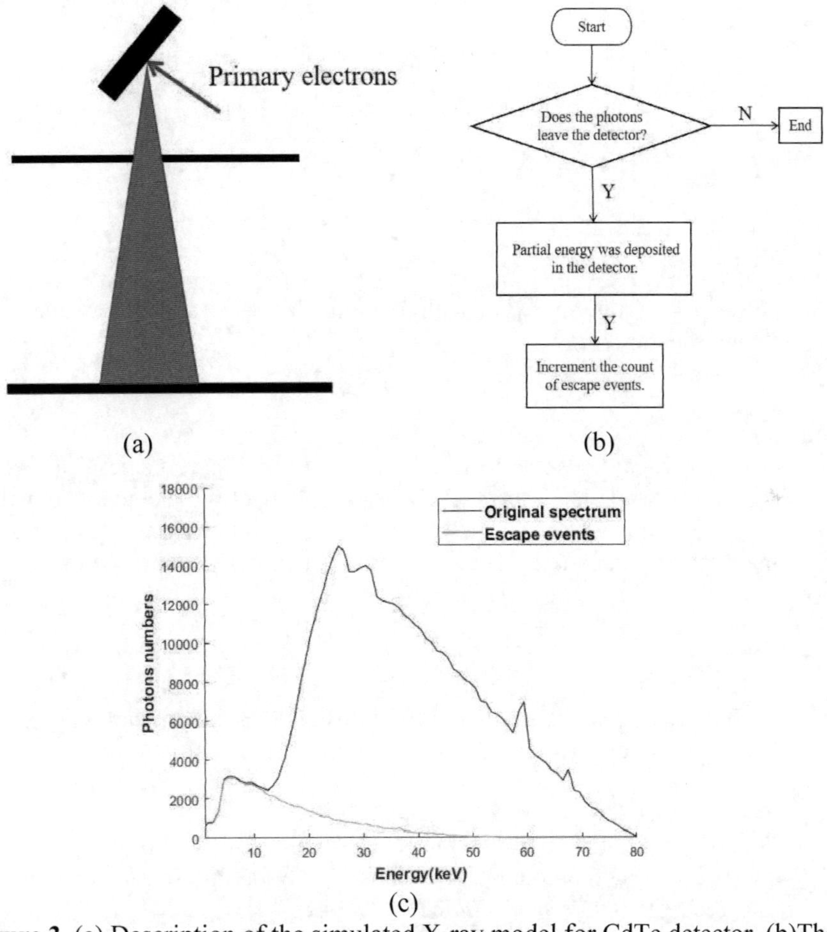

Figure 3. (a) Description of the simulated X-ray model for CdTe detector. (b)The flowchart for measuring escape events. (c) X-ray spectrum with 1 mm CdTe and the escape events counts.

3.3. Measurement of spectrum by Gd_2O_2S

In this experiment, a total filtration of 2.8 mm Al and tungsten anode angle fixed at 10° and detector thickness is 1.4 mm. Tube voltages is at 140 kVp. 2×10^9 electrons were input. The model was presented in figure 1(a).

K-edge energy of Gd is about 50.28 keV. Discontinuities can also be found in the spectrum of Gd_2O_2S. The green trace shows the escape events found in the original spectrum (figure 4).

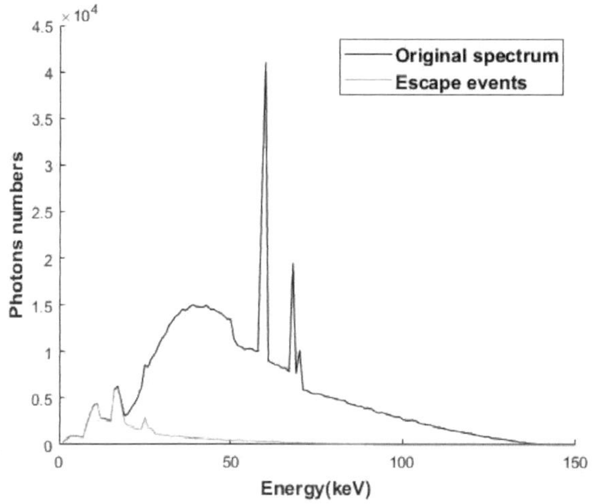

Figure 4. X-ray spectrum with 1.4 mm Gd_2O_2S and the escape events counts.

4. Conclusion

As shown in this study, K X-ray escape effects can lead to spectral inaccuracies. The experimental and simulation results matched well. The escape characteristics reflect the elements properties evidenced on the energy loss in the characteristic X-ray. The escape fraction of a detector is also strongly related to the detector material and thickness. The impact of detector's thickness will be investigated in continuous work.

References

[1] Flohr T, Ulzheimer S, Petersilka M and Schmidt B 2020 Basic principles and clinical potential of photon-counting detector CT *CJAR* **3** 19-34

[2] Huang S Y, Yang K, Abbey C K and Boone J M 2012 A semiempirical linear model of indirect, flat-panel x-ray detectors *Med. Phys.* **39** 2108-2118

[3] Leng S, Bruesewitz M, Tao S, Rajendran K, Halaweish A F, Campeau N G, Fletcher J G and McCollough C H 2019 Photon-counting Detector CT: System Design and Clinical Applications of an Emerging Technology *RadioGraphics* **39** 729-743

[4] Taguchi K 2017 Energy-sensitive photon counting detector-based X-ray computed tomography *Radiol. Phys. Technol.* **10** 8-22

[5] Alivov Y, Baturin P, Le H Q, Ducote J and Molloi S 2014 Optimization of K-edge imaging for vulnerable plaques using gold nanoparticles and energy resolved photon counting detectors: a simulation study *Phys. Med. Biol.* **59** 135-152

[6] Cao Q, Sisniega A, Brehler M, Stayman J W, Yorkston J, Siewerdsen J H and Zbijewski W 2018 Modeling and evaluation of a high-resolution CMOS detector for cone-beam CT of the extremities *Med. Phys.* **45** 114-130

[7] Singh R, Wu W, Wang G and Kalra M K 2020 Artificial intelligence in image reconstruction: The change is here *Med. Phys.* 113-125

[8] Shikhaliev P M, Fritz S G and Chapman J W 2009 Photon counting multienergy x-ray imaging: effect of the characteristic x rays on detector performance *Med. Phys.* **36** 5107-5119
[9] Si-Mohamed S A, Miailhes J, Rodesch P A, Boccalini S, Lacombe H, Leitman V, Cottin V, Boussel L and Douek P 2021 Spectral Photon-Counting CT Technology in Chest Imaging *J. Clin. Med.* **10**
[10] Can C n and Bilgici S Z 2003 An investigation of x-ray escape for an HPGe detector *X-Ray Spectrometry* **32** 276-279
[11] Ohtsuchi T, Ohmori K, Tsutsui H and Baba S 1995 Evaluation of K X-ray escape and crosstalk in CdTe detectors and multi-channel detectors *IEEE T. Nucl. Sci.* **42** 179-184
[12] Boone J M, Seibert J A, Sabol J M and Tecotzky M 1999 A Monte Carlo study of x-ray fluorescence in x-ray detectors *Med. Phys.* **26** 905-916
[13] Aoki K and Koyama M 1990 Photoelectron escape from a silicon X-ray detector: Monte Carlo study *Phys. Med. Biol.* **35** 1051-1063
[14] Kandarakis I and Cavouras D 2001 Experimental and theoretical assessment of the performance of Gd_2O_2S:Tb and La_2O_2S:Tb phosphors and Gd_2O_2S:Tb–La_2O_2S:Tb mixtures for X-ray imaging *Eur. Radiol.* **11** 1083-1091
[15] Freed M, Park S and Badano A 2010 A fast, angle-dependent, analytical model of CsI detector response for optimization of 3D x-ray breast imaging systems *Med. Phys.* **37** 2593-2605
[16] Miyajima S 2003 Thin CdTe detector in diagnostic x-ray spectroscopy *Med. Phys.* **30** 771-777
[17] Redus R H, Pantazis J A, Pantazis T J, Huber A C and Cross B J 2009 Characterization of CdTe Detectors for Quantitative X-ray Spectroscopy *IEEE T. Nucl. Sci.* **56** 2524-2532
[18] Howansky A, Mishchenko A, Lubinsky A R and Zhao W 2019 Comparison of CsI:Tl and Gd_2O_2S:Tb indirect flat panel detector x-ray imaging performance in front- and back-irradiation geometries *Med. Phys.* **46** 4857-4868
[19] Olschner F, Lund J C and Stern I 1989 Monte Carlo simulation of gamma ray spectra from semiconductor detectors *IEEE T. Nucl. Sci.* **36** 1176-1179
[20] David S, Michail C, Seferis I, Valais I, Fountos G, Liaparinos P, Kandarakis I and Kalyvas N 2016 Evaluation of Gd_2O_2S:Pr granular phosphor properties for X-ray mammography imaging *J. Lumin.* **169** 706-710
[21] Aoki K and Koyama M 1998 Measurement of diagnostic x-ray spectra using a silicon photodiode *Med. Phys.* **16** 529-536
[22] Howansky A, Peng B, Lubinsky A R and Zhao W 2017 Deriving depth-dependent light escape efficiency and optical Swank factor from measured pulse height spectra of scintillators *Med. Phys.* **44** 847-860
[23] Durak R and Özdemir Y 2001 X-RAY FLUORESCENCE ESCAPE PEAKS IN Ge(Li) DETECTORS *Instrum. Sci. Technol.* **29** 185-192
[24] Idrissi A, Dourki I, El Bardouni T, El Hajjaji O, Drissi El-Bouzaidi M and Mira M 2023 G4XRTube: A Geant4-based Monte Carlo application for X-ray tube simulation *Radiat. Phys. Chem.* **207**
[25] Fewell T R 1981 Handbook of computed tomography x-ray spectra *US Department of Health and Human Services, Public Health Service, Food and Drug Administration, Bureau of Radiological Health*

AUTHOR INDEX

Abdelraouf, M. E. 127
Abukhaled, Marwan 88
Al-Ameen, Talal 7
Alporha, R 96
Ansar, Abrar 88
Ansor, N Mohamad 52
Bercasio, C J 27
Bokhari, N. H. M. 66
Cheng, Yung-Chang 60
Chitose, Makoto 21
Diego, Ky M 109
Dikole, R. 102
Domingo, Zane Nikia C. 79
Elsaid, A. 127
Elsaid, A 14
Francis, A C 14
Gabayno, Jl F 115
Gammag, R 27, 96
Gong, Xuanyou 1
Hamidi, Z S 52
Hamidi, Z. S. 66
Hsieh, Hsiang-Yu 37
Jaruda, Ra J 115
Jitman, Somphong 73
Kandil, A. 127
Kandil, A. 14
Kijamnajsuk, Puchong 32
Kouchi, Toshinori 1, 21
Lee, Cheng-Kang 60
Li, Yiting 43
Liu, Chien-Kuan 60
Liu, Po-Liang 37
Luengviriya, Chaiya 32
Macalalad, Ernest P. 79
Mahmud, Maythem 7
Muhi, Imad 7
Niu, Luchang 43
Putungan, D B 109
Santos-Putungan, A B 109
Sese, L J 27, 96
Shariff, N N M 52
Shariff, N. N. M. 66
Shi, Zhe 134
Sibsiri, Kwanruthai 32
Su, Jian-Sheng Xie 37
Sutthiopad, Malee 32
Suzuki, Hiroki 1, 21
Wang, Dan 134
Yang, Zhi 134
Yu, Cheng-Lung 37
Zahra, W K 14
Zahra, W. K. 127
Zhang, Kun 43
Zhang, Tiangang 43